OTHER
SPACE ODYSSEYS

GREG LYNN
MICHAEL MALTZAN
ALESSANDRO POLI

OTHER
SPACE ODYSSEYS

EDITED BY
GIOVANNA BORASI
MIRKO ZARDINI

CANADIAN CENTRE FOR ARCHITECTURE
LARS MÜLLER PUBLISHERS

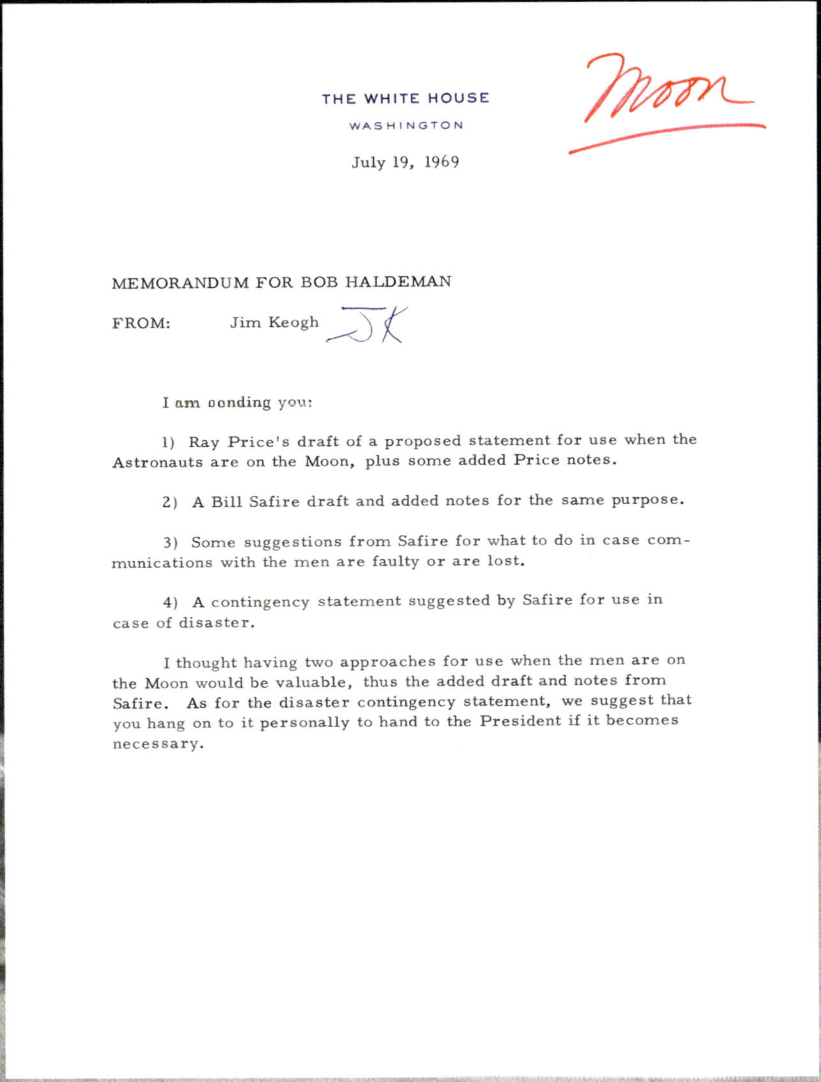

Moon

THE WHITE HOUSE

WASHINGTON

July 19, 1969

MEMORANDUM FOR BOB HALDEMAN

FROM: Jim Keogh JK

I am sending you:

1) Ray Price's draft of a proposed statement for use when the Astronauts are on the Moon, plus some added Price notes.

2) A Bill Safire draft and added notes for the same purpose.

3) Some suggestions from Safire for what to do in case communications with the men are faulty or are lost.

4) A contingency statement suggested by Safire for use in case of disaster.

I thought having two approaches for use when the men are on the Moon would be valuable, thus the added draft and notes from Safire. As for the disaster contingency statement, we suggest that you hang on to it personally to hand to the President if it becomes necessary.

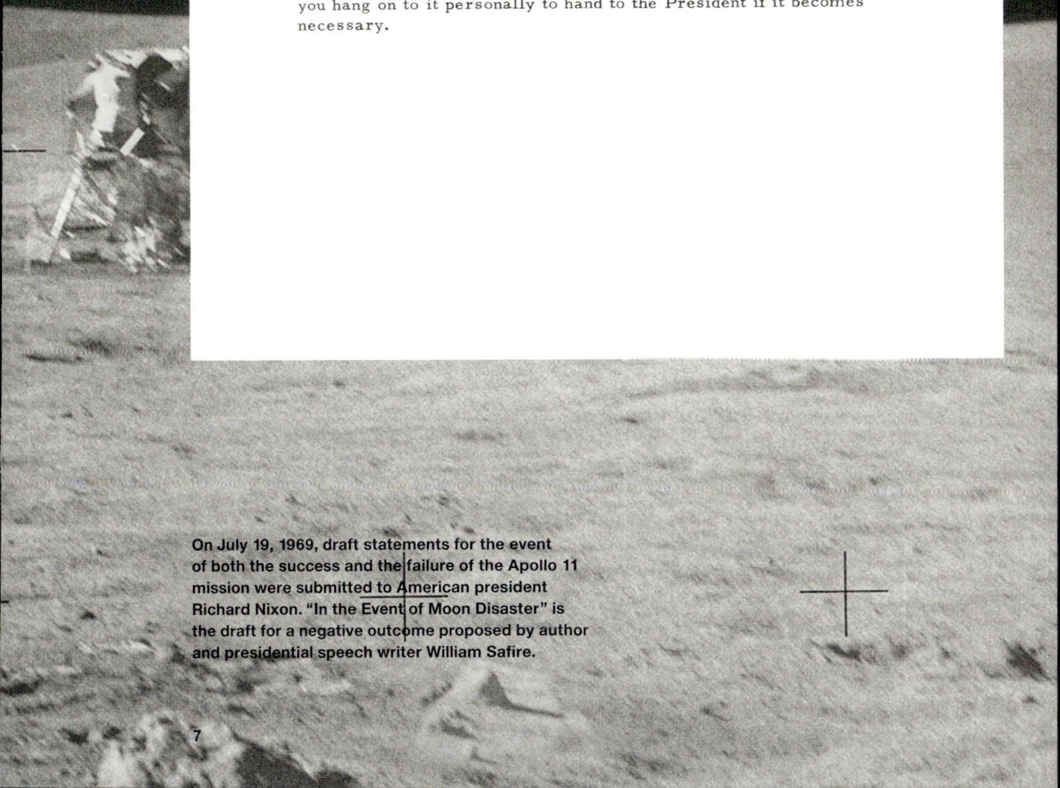

On July 19, 1969, draft statements for the event of both the success and the failure of the Apollo 11 mission were submitted to American president Richard Nixon. "In the Event of Moon Disaster" is the draft for a negative outcome proposed by author and presidential speech writer William Safire.

To   :   H. R. Haldeman

From: Bill Safire   July 18, 1969.

IN EVENT OF MOON DISASTER:

Fate has ordained that the men who went to the moon to
explore in peace will stay on the moon to rest in peace.

These brave men, Neil Armstrong and Edwin Aldrin, know
that there is no hope for their recovery. But they also know that there
is hope for mankind in their sacrifice.

These two men are laying down their lives in mankind's
most noble goal: the search for truth and understanding.

They will be mourned by their families and friends; they
will be mourned by their nation; they will be mourned by the people of
the world; they will be mourned by a Mother Earth that dared send two
of her sons into the unknown.

In their exploration, they stirred the people of the world to
feel as one; in their sacrifice, they bind more tightly the brotherhood
of man.

In ancient days, men looked at stars and saw their heroes in
the constellations. In modern times, we do much the same, but our heroes
are epic men of flesh and blood.

Others will follow, and surely find their way home. Man's search will not be denied. But these men were the first, and they will remain the foremost in our hearts.

For every human being who looks up at the moon in the nights to come will know that there is some corner of another world that is forever mankind.

PRIOR TO THE PRESIDENT'S STATEMENT:

The President should telephone each of the widows-to-be.

AFTER THE PRESIDENT'S STATEMENT, AT THE POINT WHEN NASA ENDS COMMUNICATIONS WITH THE MEN:

A clergyman should adopt the same procedure as a burial at sea, commending their souls to "the deepest of the deep," concluding with the Lord's Prayer.

VANTAGE POINTS ACCESSIBLE FOR THE FIRST TIME: A STORY IN SIX CHAPTERS

Giovanna Borasi

Chapter One: Man Will Conquer Space *Soon*

At least 125 million people in the United States alone watched man's first steps on the moon.[1] Broadcast live on July 20, 1969, this world media event was the culmination of a dream that first appeared in literature centuries ago[2] but attained popularity only in the 1950s. Space exploration offered an alluring image of the future, one that was intensely attractive in the period of economic recovery following World War II[3] and was well suited to the emerging media of television and colourful magazines. An imminent departure for other planets was a compelling prospect, and it coincided with a desire for change and widespread technological optimism.

Arthur C Clarke's non-fiction books *The Exploration of Space* (1951), which described in detail how interplanetary travel would become common in the coming decades, and *The Exploration of the Moon* (1954) found a broad audience in this atmosphere and helped to transform the idea of space travel from fiction to potential reality. Specialized magazines and digests like *Galaxy*, *Imagination*, *Astounding/Analog*, *If*, *New Worlds*, *Fantasy & Science Fiction* and *Popular Mechanics* revealed possible futures on other planets and fed belief in this "new reality."[4]

The most famous of these magazines are the issues of the American weekly *Collier's* devoted to the discovery of space. Editor Gordon Manning decided in 1951 to publish ideas and projects related to space research, and he gathered experts to publish their views. These included rocket scientist Wernher von Braun, a beneficiary of Operation Paperclip, the American program that rescued useful Nazi scientists, science fiction writer Willy Ley, illustrator Chesley Bonestell, astronomer Fred Whipple, physics professor Joseph Kaplan, doctor Heinz Haber and lawyer Oscar Schachter. Their consensus was simple: space travel was not only possible, it was an accessible idea, possible to explain and illustrate in a wealth of convincing technical detail.

The articles in the *Collier's* series, which were written by Von Braun and Ley, were captivating for their exuberantly optimistic tone expressing the certainty that a space future was imminent, but above all for the superb illustrations by artist Chesley Bonestell, which provided a colourful how-to of the conquest of space.[5] The success of this series (four million readers) ensured a television adaptation produced by Disney between 1955 and 1957 featuring the charismatic Von Braun and Ley. The series *Tomorrowland: Disney in Space and Beyond* used the nascent medium of television to project its seamless vision to an even wider audience, and captured the imaginations of millions of children.[6]

In the 1950s the boundary between everyday reality, scientific fiction, and science fiction became increasingly permeable, and figures like Von Braun reached across all three realms. His ideas for toroidal space stations, elaborated for *Collier's* in 1952, were taken up in the 1970s by NASA's Ames Research Center – under pressure to find new habitable places because of urgently felt concerns about overpopulation, the newly-discovered greenhouse effect, and fear of nuclear holocaust – and turned into designs for cylindrical, toroidal and spherical space colonies that could house ten thousand inhabitants each, freeing the human race from its confinement to a single planet.[7]

Outstanding among the many proposals made at the time for inhabitable space structures was the work of physicist Gerard O'Neill and his students at Princeton in 1969. Interested in the possibility that humans could live in space and that new technologies could be developed more effectively on non-planetary terrain, O'Neill and his students researched projects for huge pressurized cylindrical structures with near-earth gravity. These took shape in a design called the O'Neill Cylinder: the population would live on the inner surface, creating an "inside-out planet" with an environment and a landscape similar to those on earth.

The image of a future in space was clear. Now it was just a matter of making it happen.

Chapter Two: We Went to the Moon and Brought the Earth within Us

Visionary images had helped create popular support for space research and belief that it was feasible to travel to the moon, Mars, or simply roam the universe discovering new worlds and asserting the supremacy of the human race. This future came true[8] but it held a surprise.

During the Apollo 8 mission, three American astronauts (William Anders, Frank Borman, and Jim Lovell) orbited the moon in preparation for the Apollo 11 mission, which was the first to land on the moon's surface. They radioed their initial impressions of what it looked like: "Okay, Houston," Lovell said. "The moon is essentially grey. No color. Looks like plaster of Paris – or sort of a greyish beach sand."[9] From close up, the moon did not look as spectacular as the astronauts had imagined it, but something else did: they discovered the earth. "Oh my God! Look at that picture over there! Here's the earth coming up," Borman, commander of the mission, commented on the unexpected sight, "Wow, is that pretty." The photograph taken by Anders[10] on Christmas Eve 1968 captured the view of an earthrise in colour for the first time. [11] Previously, most people could only have imagined the image of the earth through the words of cosmonaut Yuri Gagarin, who said after his 1961 mission: "During the flight I saw for the first time with my own eyes the earth's spherical shape. You can see its curvature when looking to the horizon. I must say that the view of the horizon is unique and very beautiful." [12] By 1969 Buckminster Fuller already understood that unless you happened to be "a Cape Kennedy capsuler," [13] you could not have a comprehensive view of earth. Suddenly the ability to see new approaches to global problems and devise different solutions was possible by seeing the planet in its totality.

While the lunar missions of the 1960s and 1970s would make the dream of setting foot on an-other planet come true, they also gave us a better understanding of the planet we live on (for now). "The first full colour photograph of the earth from space, 22727, achieved an unprecedented iconic status precisely because of the idea it conveyed of a unified planet. In 22727 the earth is beautiful … Our singular blue planet, floating alone in … space, appears both fragile and worth saving. It is for this reason that 22727 has been credited with launching the environmental movement." [14] This shift in world view might be what Alessandro Poli was getting at when he wrote to Adolfo Natalini in 1970: "Dear Adolfo, we saw the conquest of space on TV, the greatest media event of the twentieth century. After the landing on the Moon, architecture can no longer be as we had thought-imagined-built it in our visions … no clouds, no wind, no gravity, no conflicts, no sounds, and yet the tremendous strain of discovering that we are small, because even huge monuments, solid architecture, seem so far away that they vanish." [15] The discovery of a "new reality" had opened another perspective to Poli. One could no longer think of architecture in the same way, or rather architecture as we understood it on earth could not exist out there. Projects took on another scale; they acquired new surroundings and required new tools. "Interplanetary architecture is landscape architecture … where we can live engulfed in the architecture of space." [16]

Poli also understood the necessity for new questions and new methods. How would people inhabit the moon and other planets? The magazines of the 1950s had created an idea of the future based on scientific knowledge elaborated through the imaginations of their illustrators. Reyner Banham explained in a 1961 lecture entitled "The History of the Immediate Future" how the future could be determined by plotting a curve algebraically. [17] Transferring the facts of the present to a graph, it is possible to connect them into a curve and trace this curve to the last certain point, imagining where it goes from there. For Banham, this was how to understand the future: extrapolate the lines of present tendencies.

In a similar way, Alessandro Poli, working with the other members of Superstudio on *Architettura interplanetaria* (Interplanetary architecture), [18] asks us to imagine a combinatory world, made of real and assembled images, to construct an "immediate future" made out of the present. The photomontages, storyboard and film of *Architettura interplanetaria* emphasized the idea of voyage, of an odyssey that brings us to a new territory. Running along the lunar surface, astronaut Edwin "Buzz" Aldrin comes upon two young people and they rejoice together at this chance encounter amidst new surroundings. In another image, Aldrin looks out over the Sea of Tranquility (which has become a real sea), with lights and sailboats visible in the distance.

But where are we? Poli created these photomontages by combining images of the terrestrial world with the harshness of new territory in those "cold, empty spaces of the universe." [19] Perhaps the most fascinating aspect of this future is how it is constructed: photos clipped from newspapers and illustrated Italian weeklies are combined to create a world from what we already know, and a future we have just found. The scale of this new world is at once human and terrifyingly large: a highway spans the distance between the earth and moon, and this coexists with elements of a new "lunar architecture" and features of continued life on earth. This lunar architecture is not based only on scientific principles and the primacy of technology but on a more complex system where dreams, nostalgia, memories and human presence all play fundamental roles.

The film's soundtrack is surprising in this regard: a mix of tribal African songs, Buddhist chanting and the cold voice of a news commentator create an unsettling atmosphere, but the result is captivating because it speaks not only of a new world but also of our planet earth. It is as if Poli wanted to test what human qualities might survive the journey and continue to exist in this strange new world.

Chapter Three: Odysseys that Reveal the Immediate Future

Today we are witnessing a renewed enthusiasm for space exploration with an increase in scientific expeditions, satellite launches, and the real possibility of space tourism. [20] These initiatives reveal our continued interest in exploring space, as well as a restored faith that we will benefit from the technology that will be developed, the discoveries that will be made, and from every aspect of incursion into space.

In this context, it is interesting to reconsider this theme in architectural thought, not merely to take part in the race to a new frontier and provide innovative architectural solutions for space stations, but to understand how these possibilities can influence the idea of architecture on our own planet. How can thinking about so distant a subject lead to fresh perspectives on our reality here?

The projects of Greg Lynn, Michael Maltzan and Alessandro Poli provide different avenues for approaching this question. Though science fiction has many descriptions of encounters between earthlings and the Other, unlike the joyous meeting imagined by Poli, depictions of this moment are often concerned with the discovery of something fundamentally alien, whether it occurs on earth or on another planet. Poli, Maltzan and Lynn offer us their own images of encountering otherness; in their own ways they each describe how our understanding of our environment, our situation on earth, could change after we are exposed to a new reality. Perhaps what is most interesting about their research is the ways in which their three odysseys, virtual or real, ultimately return to tell us about our own world. Producing these projects and their related images, which reflect current scientific knowledge, is a way of testing new ideas; they become a useful tool for understanding our present.

Poli and the first environmentalists realized in the 1970s that the exploration of space and the moon landing brought about a radical change in our current situation: we now have to describe our environment in broader terms, with constant reference to earth and its beyond. Poli and Superstudio's photomontages provide us the same surprise experienced by the Apollo 8 astronauts: oddly, we find features in this new dimension (a soccer field, a lake, components of a circus, etc.) that are familiar to us from earth. In *New City*, Lynn shows how our world might be represented as a non-spherical shape derived from a virtual reality, while on earth, Maltzan carries out a project whose architectural elements accept the scale changes involved in understanding the universe.

Unlike the images from the 1950s, Poli's, Maltzan's and Lynn's reflections on the possibility of experiencing or hypothesizing another reality clarify the world we know. Each of them takes us on an odyssey and through it manages to give form to something that did not exist. But Poli is different from Maltzan and Lynn in that he doesn't suggest specific new architectural forms for the world we are about to visit. He reveals an intention to use the tools already in our hands to discover another planet and inhabit an extraterrestrial landscape. All three architects do not speak of the future but of a more complex present. Banham would call it the "immediate future," novelist J G Ballard the "visionary present." [21]

Chapter Four: Greg Lynn Leaves the Earth's Surface

Greg Lynn discusses the idea and logic of fiction in a chapter of a recent book. In the introduction he explains why he is interested in using the voice of science fiction to develop an architectural project: "The science fiction mode is speculative without being promissory: new directions, technologies, lifestyles, visions, forms, materials, atmospheres, intelligence, and sensibilities are presented as already fully realized in the cultural sense. This mode of speculative writing does not permit the distance of conventional architectural theory. Addressing a popular audience, rather than a software-specialized cadre of pseudo-scientific architects isn't a bad thing either." [22] This approach lets Lynn take us on readily understandable voyages to other worlds based on narratives or fictions that describe reality in a different way; introducing new directions, technologies and forms into the realm of what is possible.

Like O'Neill with his students in the 1960s, Lynn asks himself and his students at UCLA and at the University of Applied Arts in Vienna to consider a basic point for architecture: the notion of ground. What does it mean in architectural thinking to abandon the ground, or for the ground to cease to exist? "I think that extreme things always get interesting, just because they make you rethink the whole problem,"[23] Lynn states. Imagining a new planet for the science fiction film *Divide* or working with students with these conditions forces Lynn (and us with him) to imagine a new *everything*. And just as in a science fiction story, the life, form, size and scale of the project and the nature of "new places" must be entirely rethought. Lynn constructs a world for us to move around in, as Superstudio did with *Autostrada Terra-Luna* (*Earth-moon highway*), but using digital images. Lynn's four "planets" or N.O.A.H.s (New Outer Atmosphere Habitat) – sets for the film *Divide* that are reminiscent of early NASA speculations on space-station living – take a form and scale that take advantage of the lack of gravity. These new conditions allow the "planet" or orbital station to be conceived as something other than a sphere, and for us, the inhabitants, to no longer occupy only the exterior surface of an object but live on the inside of a planet that extends far into the dark universe. Like Superstudio's photomontages, the N.O.A.H. is "a design concept that combines architecture, technology and terrestrial nature to create a new ideal of living space – one not bound by gravity and planar surface."[24] We bring the earth along with us in this odyssey as well, for inside the planet there are open spaces, microclimates and urban structures transplanted directly from our planet. This is how Lynn resolves the encounter between these two worlds.

Once these forms are established as possible in a new context, Lynn extends them to the earth with *New City*,[25] a project that incorporates the logic and a number of analogous features from the N.O.A.H.s. The earth is no longer represented as a sphere, and one may live inside the planet; scale is finite but extends to assume various forms; and so on.

And in which "reality" is *New City* located? Lynn's reflections on potential new relationships between architecture and the ground or between architecture and the law of gravity come from a science fiction screenplay; they also emerge in a simulation of the world whose content comes from the real world but whose parameters are virtual. By deriving features from a world far from ours, *New City* becomes a new way of thinking about the planet that we live on. In this regard, *New City* is a fascinating representation of contemporary life, all the more fascinating after realizing that Lynn conceived it in such a form after "exploring space."

Chapter Five: From Pasadena, Michael Maltzan Looks at What Is Light Years Away

Michael Maltzan's new building for the Jet Propulsion Laboratory (JPL) in Pasadena, California, tackles a radically different problem from the one Greg Lynn is working on. Though JPL began designing and building spacecraft to explore other worlds in the 1960s, none of the missions to the moon or beyond to the rest of our solar system undertaken by JPL involved sending human beings, animals or living things. At JPL, man remained on earth, and space exploration took place within this scope. Their primary goal was not testing the possibility of living in space, on another planet or on a space station, even temporarily, but to survey the outer reaches of space.

How to conceive of a building for scientists who observe phenomena occurring light years away while remaining on this planet? This spatio-temporal paradox makes Maltzan's project intriguing for this discussion.

Fascination with alien worlds has led to exploratory missions in search of other realities on earth, to the Antarctic, the depths of the oceans, and the highest peaks. In 1963, French oceanographer Jacques Cousteau created an underwater human colony in the Red Sea: Conshelf II, established at the Sha'ab Rumi Reef, and consisting of two underwater dwellings. "Oceanauts" (a word coined by Cousteau to convey the idea of an exploratory mission, in this case under the sea) lived at the base for about a month.[26] There are other similar experiences that explored a remote place on earth as being like a base on another planet, a place to test research under specific environmental conditions and develop knowledge of new worlds. The problem was more complex for Maltzan: the building is on earth, but the people working within it have no scientific interest in Pasadena.

The relationship between the two realities involved, earth and space, is again critical. Maltzan's design for JPL absorbs the radical juxtaposition of these two worlds in its architectural elements. The building has to speak of both realities while having a terrestrial size, scale and use. The project accommodates its limitations; it is unable to lift off and orbit so the connection with space research is expressed more conceptually than formally, by rethinking various architectural elements.

The design evokes a planet: it can be approached and navigated from all sides, without the functional hierarchy of a traditional building. The envelope became a key element in constructing a juxtaposition between the extraterrestrial subject and the site. It was conceived as a thin membrane with openings that vary in size continuously, calling to mind the way in which vision is framed by the tools inside the building, telescopes and microscopes. The pattern of scattered apertures echos the process of zooming in and out with Google Earth to look back at ourselves and experience the effect of orbiting the planet.[27] This formal device is intended to elicit the mental elasticity of JPL's scientists. The idea that the researchers link two worlds while engaged in time-delayed work detached from their object of interest fascinated Maltzan, and he offers them a building that contains within itself the multiplicity of extra- and terrestrial scales along with ideas of temporal distance.

By participating in dialogue with another "reality," Maltzan constructs his odyssey and, like Poli and Lynn, brings elements of another world back to earth.

Chapter Six: Zeno Meets Aldrin in Riparbella

A manned space mission tends to imply a return to earth. This is how Alessandro Poli describes *Cultura materiale extraurbana* (Extra-urban material culture),[28] the research project he undertook at the Faculty of Architecture of Florence in 1974 in parallel with some members of Superstudio. A main figure of this research project is Zeno, a Tuscan peasant who had lived his whole life in the same place, the same house. Because of this, Zeno is perhaps the "only" person who can converse with Aldrin (according to Poli, the two met recently in Riparbella): both have had the experience of finding themselves inside an isolated "capsule," within a closed and protected world.

When he reflects on his research today, Poli wonders what attitude or approach might allow us to grasp our reality and to learn to know our planet. Aldrin needs a gigantic apparatus and precise, technologically sophisticated instruments tended by a large team of people to survive in his capsule. Zeno uses and reuses the same objects, applying incredible resourcefulness in making and readapting them; he is completely self-sufficient. The two experiences were also distinguished by their relationship to time: Zeno has his whole life to discover his place, and Aldrin has only a few hours to comprehend a whole new world.

Lynn, Maltzan and Poli give us new vantage points on how we inhabit the planet. Greg Lynn imagined the virtual earth of *New City* only after he had "left" the earth to design humanity's new off-planet homes. Michael Maltzan designed an earthbound building after being transported by the minds of the JPL scientists into outer space. And Poli is able to interpret Zeno's story in *Cultura materiale extraurbana* because in his own way Poli too has "gone to the moon" and brought it back within him.

1 "A Remote That Broke All the Records: Camera follows astronauts to lunar landing; next challenge is color pickup from the moon," *Broadcasting Magazine*, on Broadcasting & Cable website, originally published July 1969, posted online July 17, 2009, www.broadcastingcable.com/article/315617-A_Remote_That_Broke_All_the_Records.php?rssid=20065 (accessed February 16, 2010).

2 Examples are Cyrano de Bergerac, *Les États et empires de la Lune* (1657); Voltaire, *Micromégas* (1752); Jules Verne, *De la Terre à la Lune. Trajet direct en 97 heures 20 minutes* (1865).

3 For an in-depth analysis of the success of the science fiction genre in the postwar period, see Walter A McDougall, *The Heavens and the Earth: A Political History of the Space Age* (New York: Basic Books, 1985).

4 See James Gunn (with an introduction by Isaac Asimov), *Alternate Worlds: The Illustrated History of Science Fiction* (London: Prentice-Hall International, 1975).

5 Ley and von Braun prepared a comprehensive scheme for the missions, covering subjects from the launch of a first rocket beyond the earth's atmosphere to the construction of a space station and a thoroughly detailed description of landing on the moon and Mars.

6 The television series was in three parts: *Man in Space, Man and the Moon* and *Mars and Beyond*. For an in-depth analysis of von Braun's relationship with *Collier's* magazine and his subsequent collaboration with Disney, see Mike Wright, "The Disney-Von Braun Collaboration and Its Influence on Space Exploration," in *Inner Space/Outer Space: Humanities, Technology and the Postmodern World: Selected Papers from the 1993 Southern Humanities Conference, February 12–14, 1993*, ed. Daniel Schenker, Craig Hanks and Susan Kray, 151–60 (Huntsville, AL: Southern Humanities Press, 1993). This article is also available online on the MSFC History web page at: http://history.msfc.nasa.gov/vonbraun/disney_article.html (accessed February 16, 2010).

7 Images of these designs are available on the NASA website at: www.nas.nasa.gov/Services/Education/SpaceSettlement/70sArt/art.html.

8 Since the conclusion of the Apollo program, a number of related moon hoax accounts have been advanced by various groups and individuals claiming that Apollo astronauts did not land on the moon and that NASA deceived the public into believing the landing did occur by manufacturing, destroying and creating photos, tapes, rock samples. See: Mary Bennett, David S Percy, *Dark Moon: Apollo and the Whistle-Blowers* (Great Britain: Aulis Publishers, 1999).

9 An original recording of the conversation is available at "American Experience: Race to the Moon," on the PBS website, at www.pbs.org/wgbh/amex/moon/sfeature/sf_audio_pop_01b_qt.html and a transcription appears in Andrew Chaikin, *A Man on the Moon* (London: Penguin Books, 1994), 110.

10 On the controversy as to who actually took the photo, see Andrew Chaikin, *A Man on the Moon* (London: Penguin Books, 1994), 563. For an accurate timeline of photos during the Apollo missions see Robert Poole, *Earthrise: How Man First Saw the Earth* (Yale University Press: London and New Haven, 2008). In the chapters: "From Landscape to Planet" and "From Spaceship Earth to Mother Earth," the author discusses the impact of Apollo mission images on how we understand our planet and how these triggered the birth of the environmental movement.

11 "The most enduring legacy of these pictures was nothing to do with the Moon at all. Rather it was the beginnings of a deeper concern for the Earth's environment"; John D. Barrow, *Cosmic Imagery: Key Images in the History of Science* (London: The Bodley Head, 2008), p. 156.

12 Sean Topham, *Where's My Space Age? The Rise and Fall of Futuristic Design* (Munich, Berlin, London and New York: Prestel Verlag, 2003), 32–33.

13 R Buckminster Fuller, *Operating Manual for Spaceship Earth* (1969; repr., Baden: Lars Muller Publisher, 2008), 56.

14 Vittoria Di Palma, "Zoom: Google Earth and Global Intimacy," in *Intimate Metropolis: Urban Subjects in the Modern City*, ed. Vittoria Di Palma, Diana Periton and Marina Lathouri (London and New York: Routledge, 2009), 263. The number 22727 is the NASA designation for the photo taken December 7, 1972, by the Apollo 17 astronaut Harrison Schmitt and commonly known as the "Blue Marble." This now iconic image presented the earth in its entirety with extreme clarity and vivid colours for the first time. Volker M Welter in his 2009 CCA Study Centre seminar "Toward an History of Environmental Architecture," presented his research on the complex relationship between this image and the environmental movement.

15 A letter from Alessandro Poli to Adolfo Natalini (Superstudio, Florence), 1970, never sent. Please see pp. 84–85 in the present volume.

16 At the end of the same letter, Poli explains that pens, pencils, ink and photomontages are no longer suitable for representing this new reality: what is needed is the movement of the camera. The film on the project *Architettura interplanetaria* (1971), was the first made by Superstudio.

17 See Reyner Banham, "The History of the Immediate Future," *Journal of the Royal Institute of British Architects*, 68, no. 7 (May 1961): 256–60. See also Jonathan E. Farnham, "Pure Pop for New People: Reyner Banham, Science Fiction and History," *Lotus International* 104 (May 2000): 112–31.

18 The project *Architettura interplanetaria* was first presented in "Superstudio presenta l'Architettura interplanetaria," *Casabella* no. 364. April, (1972): 46–48.

19 See notes by Alessandro Poli, "Architettura interplanetaria: Story Board," for a lecture given at the Fondazione Targetti, Florence, June 2006. Archivio Alessandro Poli.

20 In June 2004, aircraft designer Burt Rutan launched SpaceShipOne, initiating the commercial space age. In 2010, Virgin Galactic will begin taking tourists outside the earth's atmosphere. Meanwhile, Las Vegas hotel owner Robert Bigelow is developing the first commercial space station. They and other entrepreneurs consider space "the next big thing." In addition, there are space tourists who fund scientific research just for the chance to visit space once in their lives. For an in-depth analysis of this recent phenomenon, see Michael Belfiore, *Rocketeers: How a Visionary Band of Business Leaders, Engineers, and Pilots Is Boldly Privatizing Space* (New York: Smithsonian Books, Harper Collins, 2007).

21 J G Ballard, Introduction to "The Complete Stories of J. G. Ballard," (New York: W. W. Norton & Company, 2009).

22 Greg Lynn, "Fictions," in *Greg Lynn Form*, ed. Mark Rappolt (New York: Rizzoli, 2008), 281.

23 Interview with the author, Los Angeles, November 2009 (published in the present volume, p. 41).

24 Greg Lynn, "NOAH (New Outer Atmospheric Home): Sets for the Film *Divide*," in *Greg Lynn Form*, ed. Mark Rappolt (New York: Rizzoli, 2008), 353.

25 *New City* was developed in 2008 by Greg Lynn FORM with Peter Frankfurt, Alex McDowell and Imaginary Forces for the exhibition *New Elastic Mind*, curated by Paola Antonelli at the Museum of Modern Art, New York. Lynn further developed the project for the exhibition Other Space Odysseys organized by the CCA. See p. 30 in the present volume.

26 For a detailed description and pictures of Conshelf II, see "Conshelf: Cousteau's Cutting Edge," in *Beyond the Blue* online travel/adventure magazine at: www.beyondmag.co.uk/travel/conshelf.htm (accessed February 16, 2010), or "Conshelf I, II & III," Cousteau Society website, Technology, at: www.cousteau.org/technology/conshelf (accessed February 16, 2010).
In 2009, the Italian artists Isola and Norzi initiated a project to rediscover and document this habitational experiment, which still exists under the Red Sea. (See image on Isola Norzi home page, www.isolanorzi.com).

27 For a thorough analysis of the evolution of various instruments and devices for observing the world, see Vittoria Di Palma, "Zoom: Google Earth and Global Intimacy," in *Intimate Metropolis: Urban Subjects in the Modern City,* ed. Vittoria Di Palma, Diana Periton and Marina Lathouri, 239–270 (London and New York: Routledge, 2009).

28 See *L'esperienza "Cultura materiale extraurbana"* (Prato: Catalogo Vinci, 1977); Adolfo Natalini, Alessandro Poli and Cristiano Toraldo di Francia, "Viaggio con la matita tra gli artefatti del mondo contadino," *Modo,* no. 8 (March 1978): 49–53; Adolfo Natalini, Lorenzo Netti, Alessandro Poli, Cristiano Toraldo di Francia, *Cultura materiale extraurbana* (Florence: Alinea, 1983); and more recently Peter Lang and William Menking, *Superstudio: Life Without Objects* (Milan: Skira, 2003).

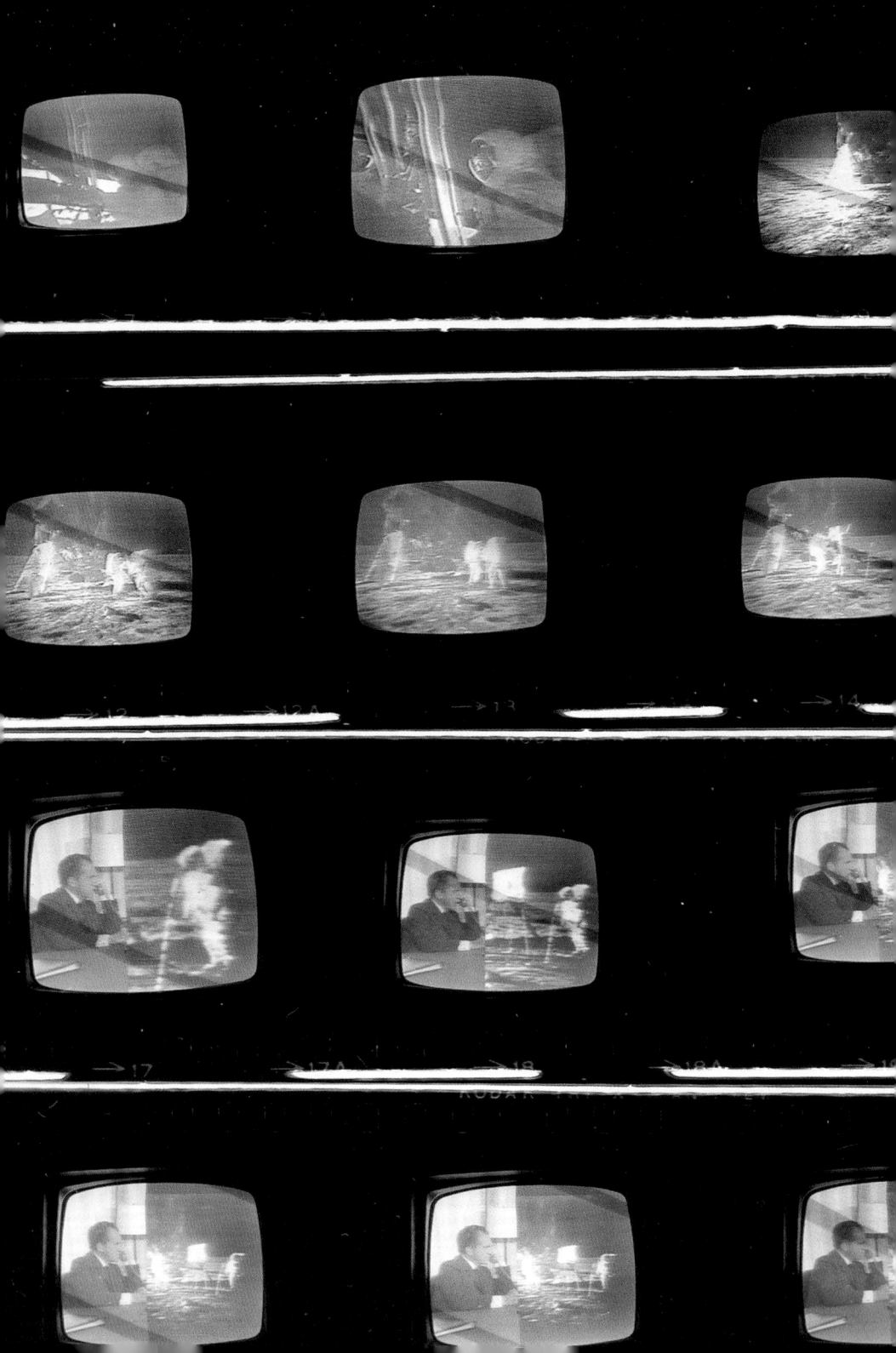

PRESIDENT RICHARD NIXON'S DAILY DIARY
(See Travel Record for Travel Activity)

PLACE DAY BEGAN	DATE (Mo., Day, Yr.)
THE WHITE HOUSE - Washington, D. C.	JULY 20, 1969

TIME   DAY
7:41 pm SUNDAY

TIME		PHONE P=Placed R=Received		ACTIVITY
In	Out	Lo	LD	
7:41	7:48	P		The President talked with his Counsel, John D. Ehrlichman.
7:53				The President went to the Residence.
8:06	8:08	P		The President talked with Secretary of State William P. Rogers.
8:19	8:22			The President met with his Assistant Military Aide, Col. Vernon C. Coffey.
8:39	8:41	P		The President talked with his Counsel, John D. Ehrlichman.
8:49				The President went to the office of his Assistant, H. R. Haldeman. Among those present were: Frank Borman, Astronaut Ronald L. Ziegler, Press Secretary H. R. Haldeman, Assistant Dwight L. Chapin, Special Assistant
9:42				The President went to his office.
9:47	9:48		P	The President talked long distance with Ernest Randall at the Space Flight Center, Houston, Texas.
9:57				The President went to the office of his Assistant, H. R. Haldeman.
11:21				The President went to his office.
11:45	11:50	P		The President held an interplanetary conversation with Apollo 11 Astronauts, Neil Armstrong and Edwin Aldrin on the Moon.
11:59				The President went to the Residence accompanied by: Patricia Nixon Julia and David Eisenhower.

U.S. GOVERNMENT PRINTING OFFICE : 1969—O-332-068

Page 3 of 3 Page(s).

US president Nixon's daily diary from July 20,
1969, notes an interplanetary phone call to Apollo 11
astronauts on the moon's surface. The phone call,
placed from the White House telephone at 11:45 pm,
lasted approximately 5 min.

FOR ME, GRAVITY
RELATIVE TERM,
MOVEMENT, ORBIT
STABILITY BETWE

GREG LYNN

S A VERY
SCRIBING
AND DYNAMIC
N OBJECTS.

GREG LYNN (born in 1964 in North Olmsted, Ohio) is the principal of Greg Lynn FORM, based in Los Angeles. He holds both a Bachelor of Environmental Design and a Bachelor of Philosophy from Miami University of Ohio, and he received a Master of Architecture degree from Princeton University.

Greg Lynn FORM has been at the cutting edge of computer-aided design in the field of architecture. His projects, teachings and writings have been influential in the use of advanced technology for design and fabrication.

His work on the *Embryological House* (1997–2001), a ground-breaking early work of digitally created theoretical architecture, is held in the permanent collection of the CCA.

N.O.A.H. (New Outer Atmospheric Habitat)

Greg Lynn

When Jörg Tittel and Ethan Ryker were developing *Divide*, a science fiction film addressing some of the socio-political mores that shape our world and future, he invited me to design one of the characters: four new planets. The N.O.A.H. (New Outer Atmospheric Habitat) structures are internally porous, city-scaled, man-made space stations housing millions. They are designed to take advan-tage of orientations unique to the absence of gravity, much like the interior organization and structures of cells or bacteria. The N.O.A.H.s are riddled with open spaces and microclimates directly transplanted from earth. It is a design concept combining architecture, technology, and terrestrial nature to create a new ideal of living space – one not bound by gravity or planar sur-faces. From a distance, a N.O.A.H. resembles a single vast discrete shape, but on closer observation it becomes an amalgamation of mutable and modular cellular pockets. These cells create a variety of structural layers – similar to a coral reef – and read as chambers and volumes within. When intersected with the outer skin, they create crater-like openings that allow for the circulation of light and air throughout the interior spaces.

N.O.A.H. (New Outer Atmospheric Habitat),
Greg Lynn, 2004
Asia, one of the "planets" designed for the film
Divide: the structure is riddled with open spaces
and microclimates directly transplanted from
earth's continent of the same name.
Model: 3-D printed ABS plastic
Pages 26–29: Side, front and axonometric views
of the Asia planet.
Rendered frames

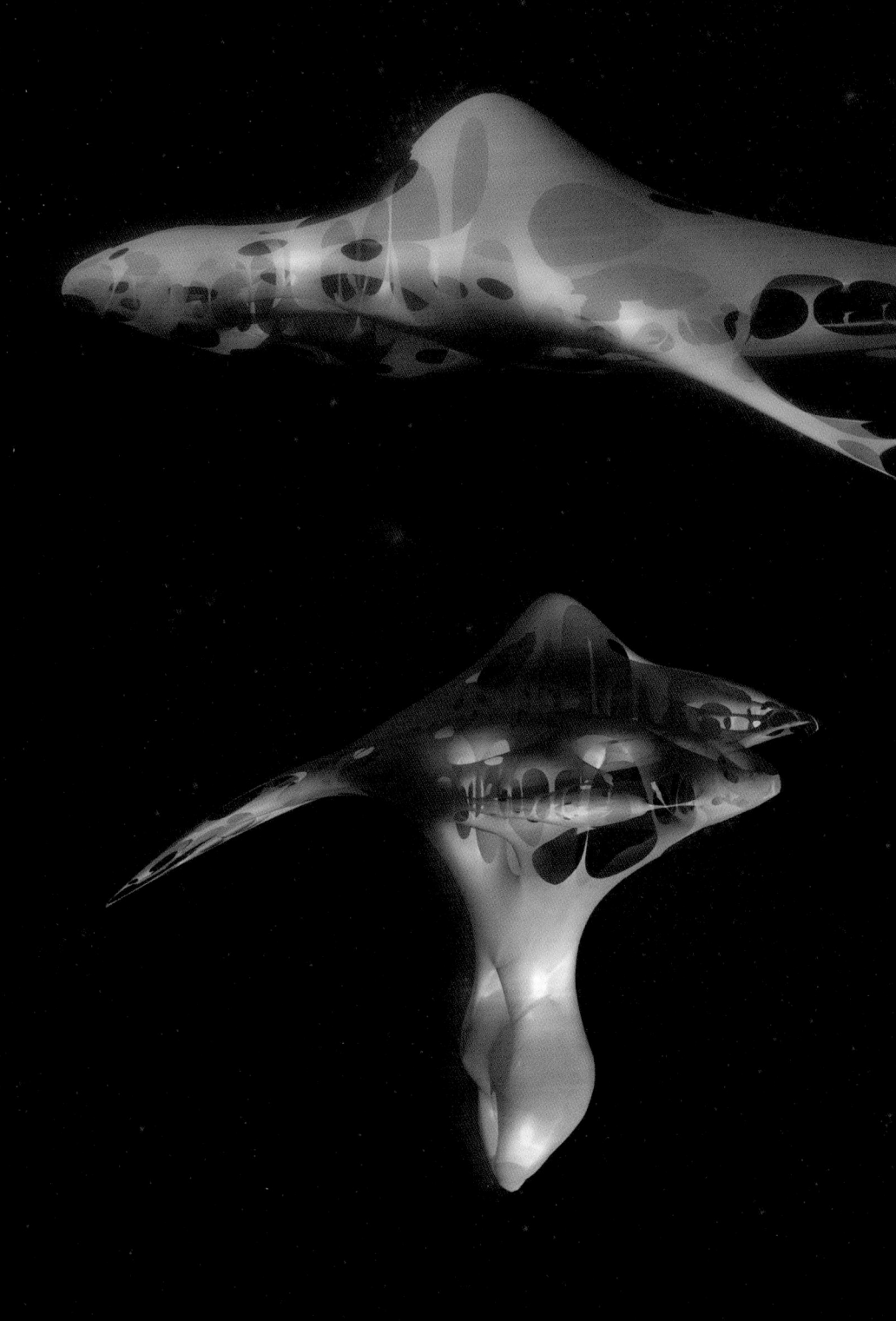

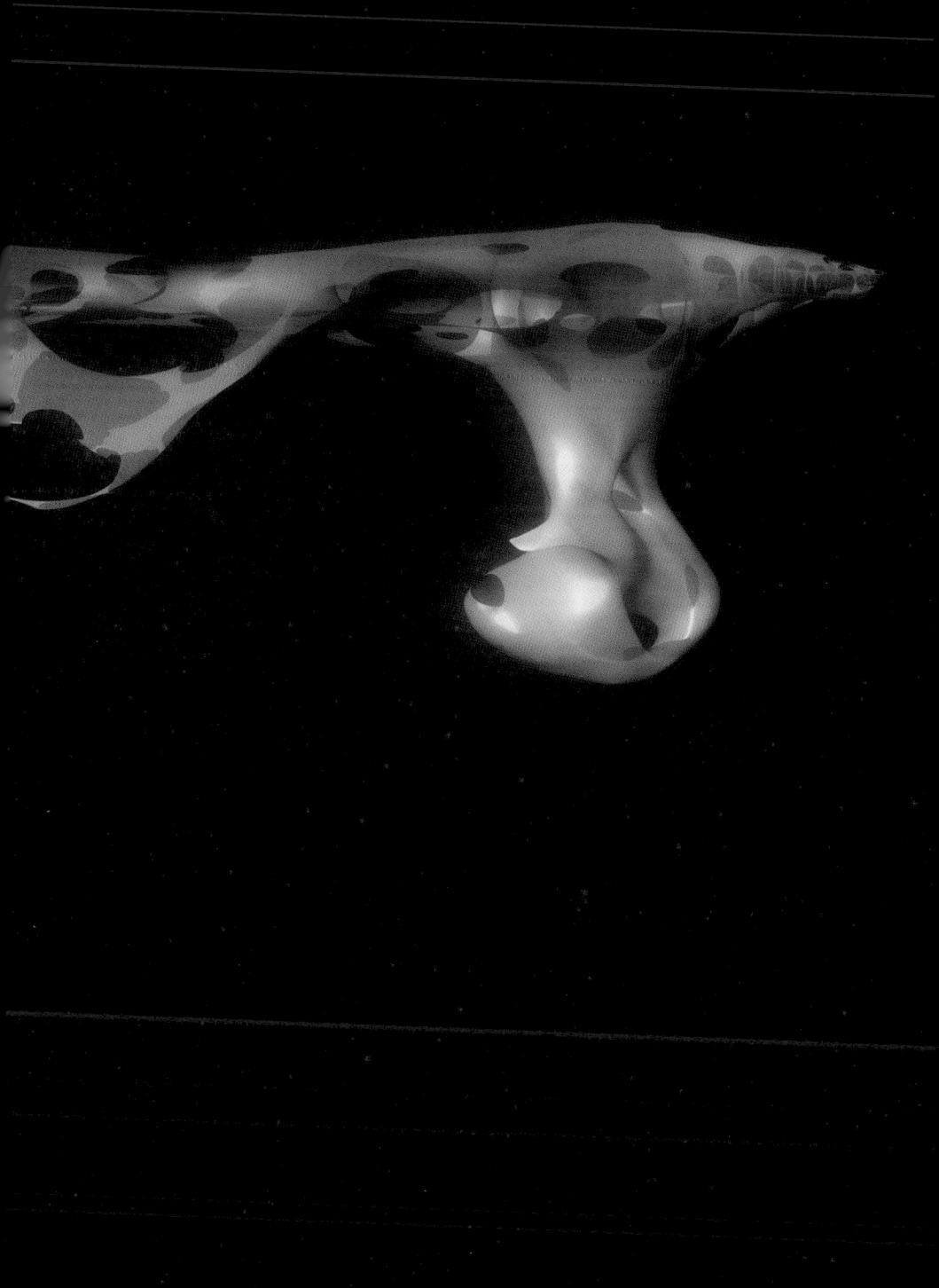

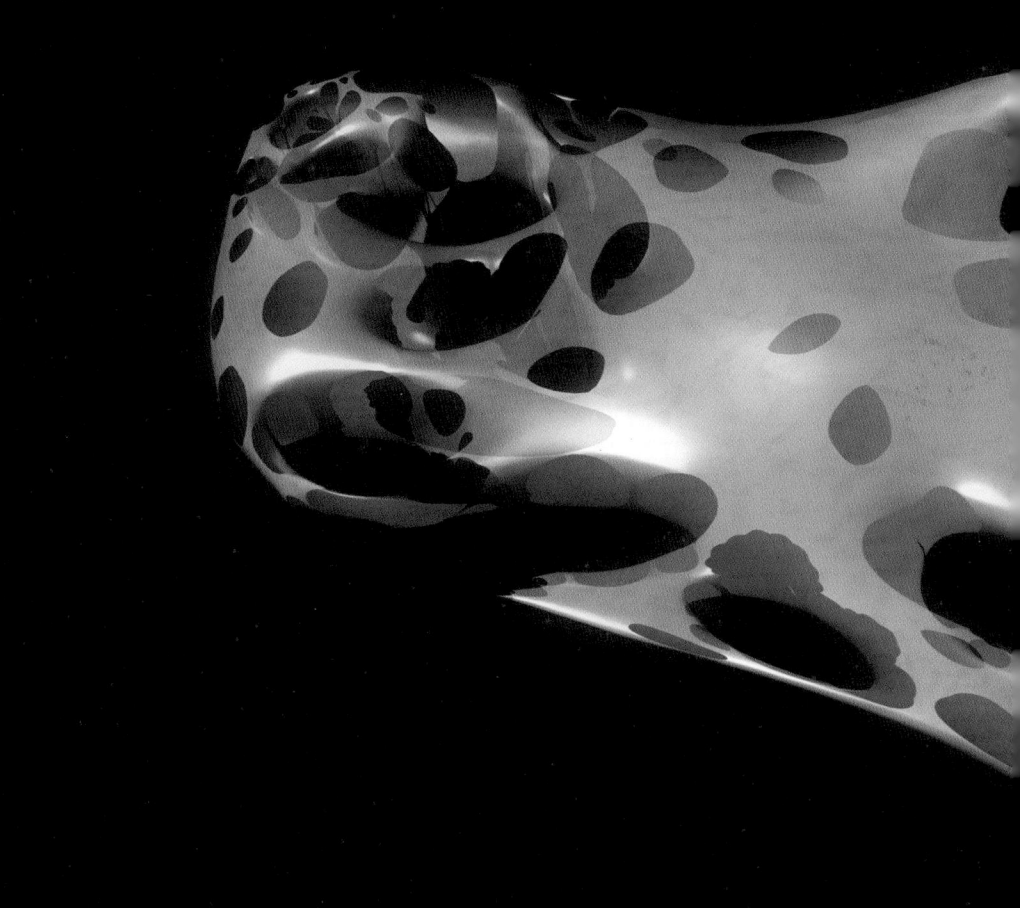

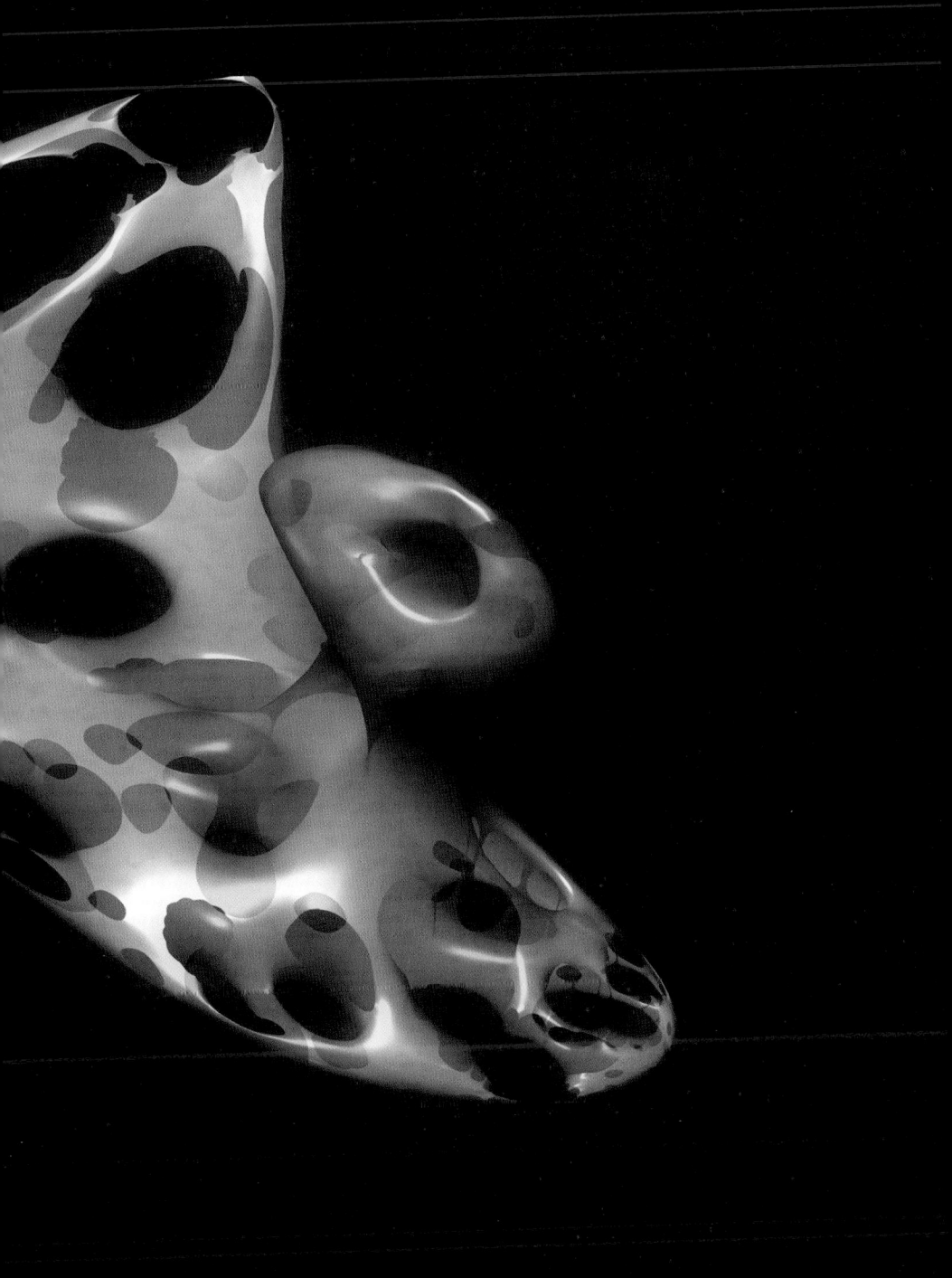

NEW CITY

Greg Lynn

New City is a concept for a living virtual world that is parallel and simultaneous to ours. A collaboration of Hollywood producer Peter Frankfurt, Greg Lynn, Hollywood production designer Alex McDowell, Imaginary Forces and Greg Lynn FORM, the design of this new virtual space marks the development of the first architecturally considered virtual world. Online social networking has had a huge impact on the way we communicate now, and this is really just the beginning; it only scratches the surface of the way we're going to interact with each other, communicate and experience the world.

Our understanding of space is always changing. We once thought the world was flat because we drew the world as flat. All kinds of discoveries gave the world shape and depth. The way we conceptualize a world as a sphere changed the way we thought about space and distance.

We decided to think of *New City* as the entire world, so the world is for the first time one connected city with 8 billion addresses, all located in a single virtual destination. It wasn't just about information design and geography; it was about how you'd experience this world.

New City is the first virtual place where architectural investigation can happen. Thinking about what the space of this city would be, we obviously didn't want to work with a flat map. We also didn't want to work with an earth or a globe. So, we thought of *New City* as a series of manifolds, which are reconfigurable pieces of geometry that can distort, fold through themselves, and connect in ways that they couldn't using a plane or a sphere. We decided to make each of the continents on the earth into one of these manifolds, and then let them move and ebb and flow with different kinds of communication and information.

What you would see if you went to *New City* is based on what you want to see. It's a series of lenses of information that you can overlay on the space to experience the city in very different ways. From abstract charts and data, you can move seamlessly into the three-dimensional space of a virtual city.

New City, Greg Lynn, 2008–
Right and pages 32–33: Views of *New City*,
a virtual world that is parallel and simultaneous
to ours. Each continent is represented as
a reconfigurable piece of geometry, a manifold.
Rendered frames

Pages 34–35: Details of different urban textures
inside the manifolds of *New City*. The length
and pattern of the filaments inside the curves
represent the density of a location. On the *left*
is Europe; on the *right* is North America.
Models: 3-D printed ABS plastic

Collier's

March 22, 1952 ● Fifteen Cents

Man Will Conquer Space Soon

TOP SCIENTISTS TELL HOW IN 15 STARTLING PAGES

GREG LYNN AND GIOVANNA BORASI MEET IN VENICE, LOS ANGELES, IN THE FALL OF 2009

Giovanna Borasi: Where were you on the 20th of July 1969? Were you, like many others, in front of the TV to see the broadcast of the first men on the moon? Greg Lynn: **In July of '69 I would have been four and a half years old. So I was little. And I was in Ohio. I don't remember the time of day, but I do remember watching it on television in black and white. I also remember watching more than the landing on the moon. I remember the launch and everything. To a four year old, I think the launch was more exciting than the concept of getting on the moon: finding out people were leaving the earth and flying like that in a rocket. But I do remember, as a kid, always looking at the moon and thinking that was a place you could go. I know my parents never would have thought that. So, from my first memories, I've always thought of the moon almost like I would think of Hawaii. I think my chances of sailing a boat to Hawaii are maybe better than flying to the moon, but I think of it as the same *kind* of thing—that that's part of my universe. It's part of my potential habitat. Always has been.**

the early 1950s, *Collier's* magazine published a series of ticles about space exploration German scientist Wernher n Braun. These articles were e basis for a Disney 3-part cience-factual" television eries: *Man in Space, Man id the Moon,* and *Mars and eyond,* which aired between 55 and 1957.

Giovanna Borasi: And after that event were you interested in science fiction? Greg Lynn: **My interest in science fiction was always in near future science fiction. I was never into aliens and not so much even into space travel. My science fiction interests have always been about the uncanny things that could happen on earth. I would also say that the space age and the nuclear age, for my generation, were combined. So, Godzilla and nuclear and genetic mutants on earth have always been blurred together with space travel; it was all one big science fiction mess.   I remember, as a kid, drawing environments based on NASA. If I was drawing something, I probably wouldn't be drawing a house. I would be drawing a little bubble that you would live in on the moon, or a little bubble you would live in in Antarctica, or something like that. So, I associate images like Poli's collages with being a kid.**

Giovanna Borasi: In this project we are working on together, it is fascinating for me to discover in your, Alessandro Poli's and Michael Maltzan's projects the existence of a continuous and complex relationship between the earth, the moon and outer space.   So, I was wondering why you think we're so fascinated with going out there?   NASA's Mission to Planet Earth was initiated well after going to the moon. Why did we decide to invest in the journey to space before really understanding our own planet? Maybe it's a fascination we could find in some of your projects. It's a fascination with expanding knowledge, expanding territory, discovering something, bringing it back, in order to understand

On July 20, 1969, 125 million people in the United States alone watched as Neil Armstrong (followed by Buzz Aldrin) became the first man on the moon.

People gathered to watch this world media event on black and white home televisions or in public spaces, such as Central Park.

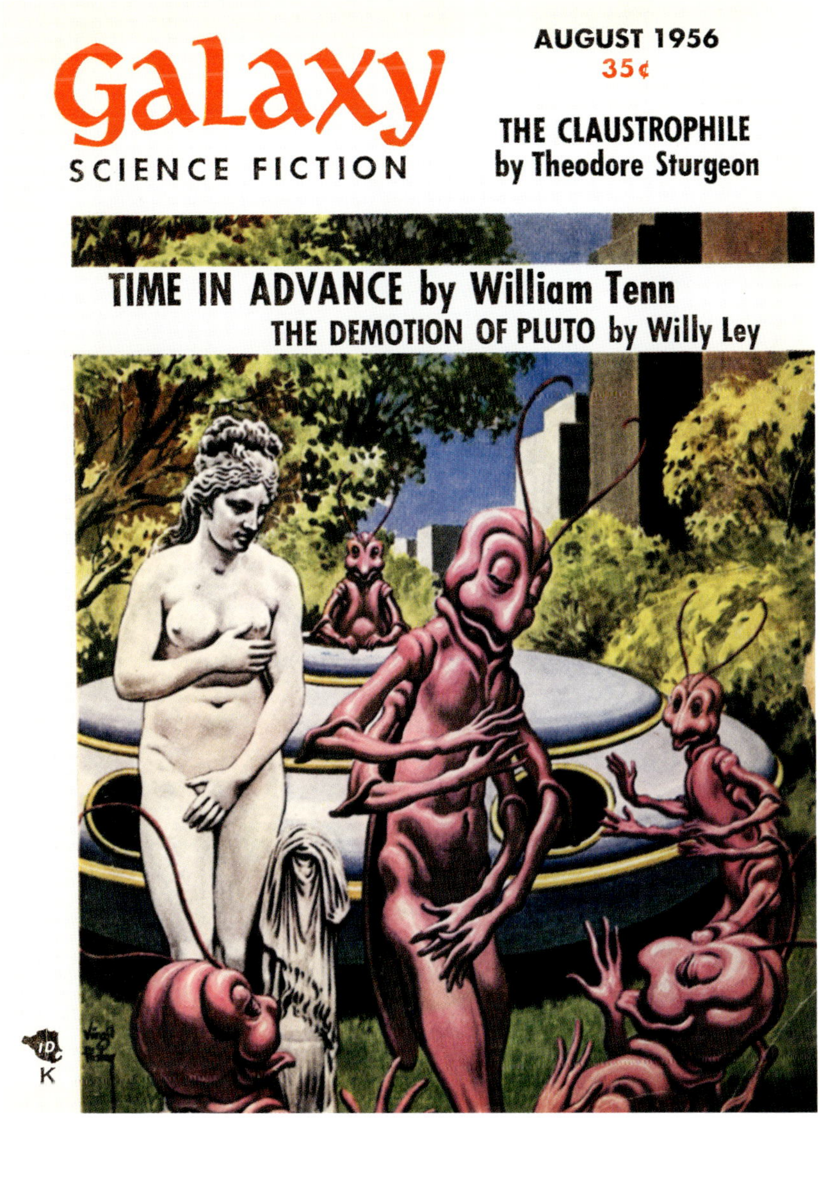

Galaxy
SCIENCE FICTION

AUGUST 1956
35¢

THE CLAUSTROPHILE
by Theodore Sturgeon

TIME IN ADVANCE by William Tenn
THE DEMOTION OF PLUTO by Willy Ley

Galaxy was an American science fiction digest published from 1950 to 1980. The brainchild of editor HJ Gold, it featured stories by well-known authors like Isaac Asimov, space art by illustrator Chesley Bonestell, and a column called "For Your Information" by scientist Willy Ley.

better our own place? Greg Lynn: **Well, for me, the space program in the '60s really represented manifest destiny, and exploration, and adventure and all that stuff, very much so. The extraterrestrial quality is also something that I think is a very big deal. That's the strangeness of something that's not of this world, but is so adjacent that it feels natural. It's like the little Martians in the movies, they're enough like humans that you feel a strange relation with them. Roboticists and renderers or animators talk about the "uncanny valley," which describes the nearly lifelike quality that is familiar enough to not be artificial but is still not-quite-right; and how creepy it makes us feel. Of course for robot designers and renderers this strange quality is to be avoided if the ambition is for familiarity and normality. I think the extraterrestrial project is really about something that's close enough to familiar but different enough that it's strange and exotic. Architecturally, that's always what I go for. I want something that looks within a historical lineage, but is different enough that it looks a little bit novel. I am interested in the uncanny valley as a positive quality.**

Giovanna Borasi: I'm wondering if by placing your projects in outer space or in a virtual world, you're giving yourself a new frame of reference where you can experiment and imagine a different architecture under very different conditions.

Greg Lynn: **Well, as a young writer and lecturer, I used to make a big point all the time about gravity. I remember at one of the Any Conferences, I responded to either Léon Krier or Rafael Moneo who said something like, "One thing that never changes is gravity, and so we don't ever change anything because gravity is our biggest concern in architecture." I remarked that gravity was not a fact but a continuously debated theory and that there were conferences much like the one we were attending where scientists discussed the concept of gravity. I understand that for someone like Rafael Moneo, who learned about gravity as a kid without the existence of a space mission, that it meant things falling off the table onto the floor. For me gravity was about vectors, orbits, velocity and slingshot curves around the moon on the way to Mars. I remember as a kid learning that gravity was a concept describing the attraction of masses, and that gravity was what drew things to the earth, but it was also what made things orbit in curves. I remember seeing astronauts bouncing around up there. And so for me gravity was a very relative term, and it had more to do with movement, and orbits, and dynamic stability** between objects and things like that.   And so a conventional thing like gravity thought of in a different scale or in a different context gets really interesting. Instead of columns being all vertical, you could suddenly think of shells and all kinds of different structures just by changing the frame of how you look at it.

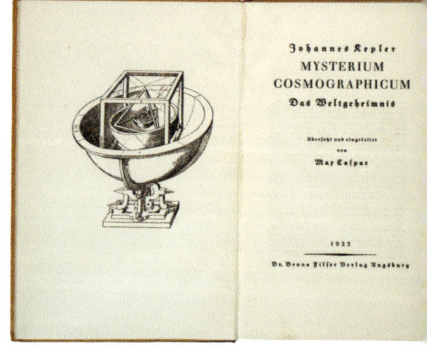

In *Mysterium Cosmographicum*, published in 1596, German astronomer Johannes Kepler proposed that the distance relationships between the planets known at the time could be understood in terms of the 5 Platonic solids enclosed within a sphere that represented the orbit of Saturn.

I think the space age did that for a lot of people. You could say there were people whose world view was formed before the moon mission and others whose ideas were formed afterwards.

Giovanna Borasi: Interesting. So, we could say that *Other Space Odysseys* is about how that event affected people's ways of thinking about architecture.

ernher von Braun's vision of a space station
at would create its own gravity by spinning was
istrated by Chesley Bonestell and published by
ollier's magazine in 1952. It was conceived
function as a navigational aid, meteorological
ation, military platform and way station for
ace exploration. Von Braun's design for a space
ation was bypassed in the 1960s when NASA
lopted the Apollo program.

If you look at the cover of the *Whole Earth Catalog* you understand that that trip really changed the way we look at the earth. This makes me think of *New City*: you describe it as a "real virtual place." Did you have a similar intention – that is, bringing everybody on a journey and producing the effect of looking at our reality in a different way?   Greg Lynn: **Sure. Well, I think that when we started *New City*, the idea was to do an ideal city that had a new system of relations based on a new paradigm of geometry and space. Many of the ideal cities proposed hundreds of years ago were all about geometries of defence, trade and circulation.   The first book assigned to my freshman class in college was by James Lovelock – the guy who invented Gaia theory, as well as being a scientist involved in the Martian probes. Lovelock talked about seeing the earth from space, you realized that the planet was alive, whereas you had always thought of it as a piece of geometry. We had the same ambition for *New City* – that it would be defined by new geometries but that it would be a complex interconnected organism. Working with Alex McDowell and Peter Frankfurt, we tried to bring this liveliness and animation through the use of media to make the geometries change and come alive, in the same way that everybody saw the earth as an organism. This ambition must be because of those first photographs from space – the big blue marble images. I was born in a time where the vision of the earth is as a discrete round animal sphere hurtling through space, surrounded by dynamic atmosphere. I don't think anybody ever really thought of it that way until they saw those pictures.**

Giovanna Borasi: The first photo of the earthrise taken by man from space was published in a newspaper the wrong way. The image was published horizontally, as though a sunrise, with a distant earth visible above the horizon of the moon. At that time no one could imagine it in any other way.   The first time we met you were speaking about these guys (Richard Branson, Robert Bigelow, Burt Rutan) who are realizing the project of space tourism and it seems it will happen soon. Greg Lynn: **They just announced last week the first hotel from Spain or something, right?**

Giovanna Borasi: So it is becoming real? A real opportunity to build something that until now has been the domain of science fiction, or developed as movie scenery or set. Do you think this will become interesting for architecture? Greg Lynn: **Well, I think that extreme things always get interesting, just because they make you rethink the whole problem. Having a hotel in the outer**

DE LA TERRE A LA LUNE

atmosphere raises all kinds of concerns like transportation: should it be inflated or flat-packed? Flat-packing is a more interesting problem than, let's say, even how much something weighs. How compact can you get it and how big can you unfold it? This is an interesting problem that comes from space colonization but has repercussions in everyday design. Or how you think about the ground, like we did with the Vienna students: you'll want some gravity, and how do you generate it? With those toroids? Wernher von Braun did that, figured out how much gravity we'd get by spinning a dough-

his 1867 novel *From the Earth the Moon* (with illustrations y Henri de Montaut), Jules Verne nticipated many details of the pollo program in his account f a 3-person mission launched om "Tampa Town," Florida.

nut. All kinds of problems like that are new problems because of an extreme environment. But then it makes you think about everything else totally differently. So, if you think about flat-packing an environment into a shuttle bay, suddenly you think about flat-packing things to put on the back of a tractor trailer. So, yeah, I think all kinds of things will come out of that. At a cultural level, what will it mean to have people going to and staying in space? For somebody to become an astronaut as a vacation will really change the way people think about space, and distance, and communication, and travel and all that stuff.

Giovanna Borasi: Do you like or despise the term "future"? Greg Lynn: I'm always suspicious of futurists. There was an industry – there still kind of is an industry – of futurists. I think of them very much like the industry of, you know, people that predict what smells or colours or whatever things should be. It works to a certain extent. It's also just not that interesting. Near future I like better than future. But, like I said, I always took it as a given that space travel, outer atmosphere, trips to the moon, all that stuff would happen in my lifetime. I never believed that I would commute to work with a jet-pack, but I would say flying to the moon: *that I* understood as part of the realm of possibility.

Giovanna Borasi: Do you think that your projects are related to technology and to the knowledge that we have in a very specific moment, and could you imagine that *New City* in another moment would *be even pushed further*? Greg Lynn: Well, it's funny. You asked where I was during the moon landing and whether I was watching television: it was as much about the television as it was about going into outer space. I mean the feeling of community. I'm sure there were other events which everybody watched – like the Beatles on Ed Sullivan – that made a global community around technology. But that was definitely a big one. And so I do think *New City* is about making a telecommunication space that has a collective feel to it, as much as anything. One thing about communication right now, and even just entertainment, in terms of media, is that there's no collective "there," the way there was before. I've never been a fan of interactivity, because usually what that means is either the deferral of community in favour of individuality or the deferral of design in favour of market selection. So, people will say, "We're gonna put some piece of technology on the web and it'll be interactive," and you'll say, "Well, what's it

supposed to do?" And they'll say, "Well, the user will decide." I find this deferral of design and community in favour of individual customization at best boring and at worst anti-social. So, I do think that there is something about the landing on the moon that I miss today – a positive human event that you can put a date and a time on. The closest thing to the moon landing and rocket launches would certainly be 9/11, where everybody knows where they were, and watched the TV at the same time, and it was a global event of that magnitude. I suppose it is not an accident that two of the three of us that worked together on the World Trade Center Site Memorial Competition cooked up *New City* as a collaboration.

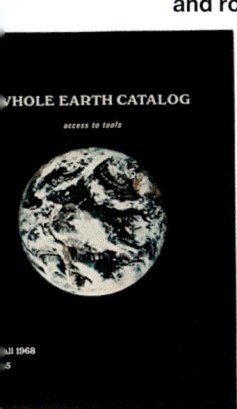

Giovanna Borasi: Going back to this idea of relationship between technology and a project: when they asked Jules Verne how he could imagine all these kinds of things, he said "I use physics. It invents." Greg Lynn: **That's interesting. That's what Jules Verne said?**

Giovanna Borasi: Yeah. Greg Lynn: **Wow.**

Giovanna Borasi: So, I don't know if you could actually say the same for you. Do new tools, technologies and media inspire you, or do you have ideas and then search for the tools to realize them? Greg Lynn: **I wouldn't say science is my dominant narrative... I would never make claims to scientific knowledge. I might claim geometric and mathematical inspiration. So, if I had to say what was the equivalent of "I use physics" for me, it would be geometry. I'm in a phase of my life where I'm thinking a lot about... exploration might be the wrong term. In the '60s and '70s, I spent a lot of my time charting courses and trajectories because I did a lot of sailing. I learned how to use a sextant, and do celestial navigation, and stuff like that. I remember thinking a lot about geometry in terms of pathways, movements, and trajectories, and so I've always thought of geometry, not in terms of a cube, and a sphere and pyramid – like primitive Platonic geome-tries – but I've always thought of geometries as arcs, and vectors, and parabolas, and curves moving in space. I've always thought of gravity as orbits, rather than gravity as a plane.   So, yeah, I would say on that level that geometry has some relationship to physics, yes. Forms in space and motion have always fascinated me. As for other technologies and media, I always collaborate.**

Giovanna Borasi: Poli explained to me that the first Superstudio movie for *Architettura interplanetaria*, was made with the belief that you had to use different tools to visualize unknown things. For him the project design happened in the drafting of the storyboards and was realized in the actual production of the movie. I'm sure the project developed in a certain way because its final presentation would be a movie. What was your experience with *Divide*? Greg Lynn: **Well, Poli was writing the movie, whereas in *Divide* my work was just the design of one of the central characters. Film-makers now understand that in some cases the design environment is one of the characters in a film. The whole media thing is really**

Published in 1964, *a history of rockets in space*
by Courtlandt Canby was the first volume of the *New
and Illustrated Library of Science and Innovation*,
which also included titles on the history of ships and
seafaring, astronomy, electricity, communications
and weaponry. This illustration, from Convair,
the company that manufactured the rockets used
to launch robotic missions to the moon in
preparation for the Apollo program, envisions
a "soft" manned landing on the moon.

interesting to me now, in a way that it wasn't a while ago. I used to think of media in either the terms defined by Robert Venturi and Denise Scott Brown, as something that inhabited architecture's surface, or in the terms of Mark Wigley, as a mode of presentation. Now, I think it's something much more intrinsic to space, volume and environment. Instead of starting with the question, "How do you apply media to architecture to make it livelier or more information rich?" now, I think it has a much more profound effect on spatial experience. I now will flirt with things that are narrative, but I would never be the author of a narrative architecture.

Giovanna Borasi: But how would you design a planet? Greg Lynn: **See, that's why I would defer to the script. What was great about** *Divide* **was the scale. When Jörg Tittel came to me and said, "Let me just tell you about this film we're doing; we want you to design several space colonies," I pictured the things that I would draw as a little kid, a little bubble with a few dozen people in it. And he said, "No, no, no. The whole motor for the movie is a technology for moving all the valuable masses from Earth into orbiting space colonies. What would happen if you could move mass? For free. What would you bring as a curator of the earth and how would you design it?" How would ground and the habitat for the Taj Mahal, the White House, and Central Park be designed in outer orbit? This narrative is what triggered the design for the N.O.A.H.s.**

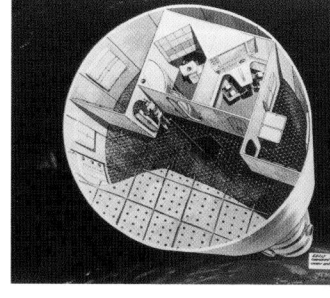

Space architecture: The Work of John Frassanito & Associates for NASA, by John Zukowsky (1999), highlights conceptual design drawings for crew quarters in the Skylab space station. The science and engineering laboratory was launched into earth orbit in 1973. Three crews of 3 men each visited the station, with their missions lasting 28, 59 and 84 days. Skylab fell from orbit on July 11, 1979.

Giovanna Borasi: So, this is also why you played with the name N.O.A.H. for the planets: they are a sort of new ark. Greg Lynn: **Yes. It's where you take all the stuff and start over. And so everybody that can, goes up there. And then all that the earth is good for is natural resources. So when Jörg was telling me all this, I was like, "Whoa, so it's not really for the people, it's actually for the Taj Mahal and Central Park. I'm designing a new place where the historic stuff goes, rather than a futuristic colony." And that was all just coming from this script. And I think** *New City* **was a direct result of that shift in thinking about scale.   I always make jokes to my friend Jeff Kipnis. He'll try to throw out a number like 50 billion something, and I'll say that doesn't actually sound like a very big number. I remember when somebody was a millionaire that was a lot of money, and then a billionaire was a lot of money. And so the same thing with distance. A million miles used to seem like a lot. But now, actually, that doesn't seem like so much. I have flown more than two million miles in my lifetime (or so my frequent** *flyer* **cards tell me).**

Giovanna Borasi: In the concept for *Divide* an important point is the idea that people would no longer live on the surface of the planet. Greg Lynn: **Yeah. When you get to a certain scale, like with the earth, what seems to be a line starts to become a bowl. And that's a very tricky thing. As we were looking**

In the 1970s NASA Ames Research Center conducted summer studies, led by Princeton physicist and space colony designer Gerard O'Neill, to imagine space colonies that could house about 10 000 people each. These artistic renderings by Don Davis and Rick Guidice visualize the proposed toroidal and cylindrical colonies.

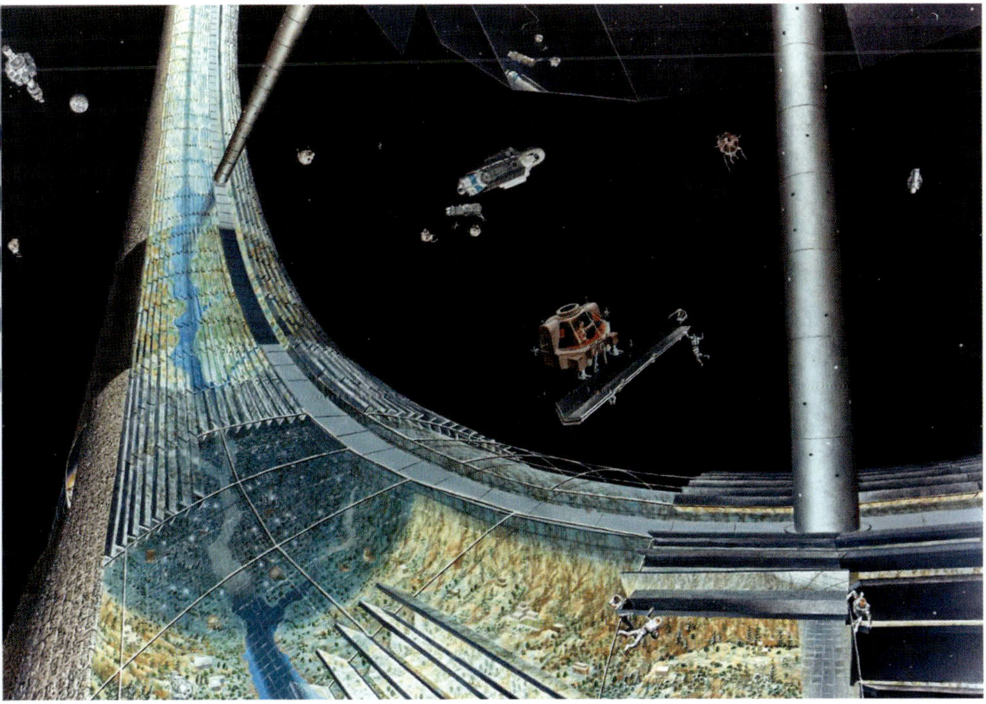

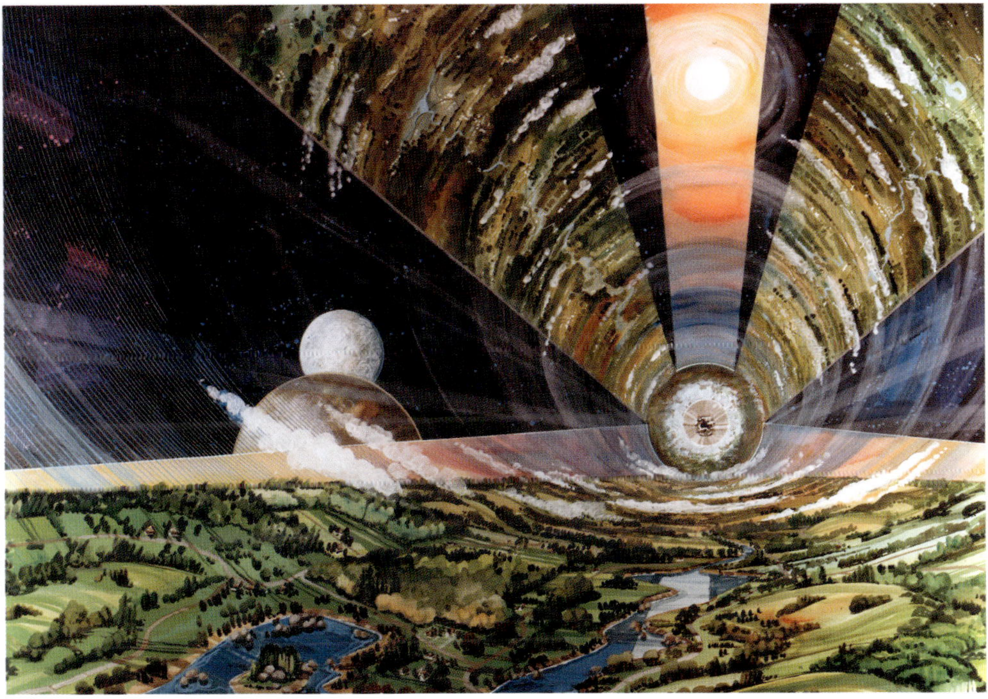

at designing urban fabric on the manifolds of *New City*'s ground, months were spent scaling building heights so that you get a sense of a curved horizon without feeling like the city is in a closed bubble. I don't want *New City* to get so huge that the sense of a space is lost but I also don't want it to get so tight that you sense you are in a bubble. What you were saying about those pictures from the moon looking back at the earth, and from the shuttle looking back: the whole great thing is finding out that everybody's living on the surface over your head, or that it's in the round.   And Superstudio always plays that game with a big horizon landing on something that doesn't really accept the horizon so well.

Six models of moon rocks from the Apollo 11, 15, 16 and 17 missions with descriptive captions are housed in a Plexiglas display case and available for loan from the NASA Exhibits Program, Johnson Space Center, Houston, TX.

Giovanna Borasi: Why did you name your project *New City*? Is this choice related to the scale that it will have? The fact that it's defined as a city suggests to me the idea that it's not endless or infinite, but maybe it could always grow or evolve into something else that we could not describe yet with the parameters that we know now. Greg Lynn: **The idea is that it's got an urban address for every person on earth. That was the first premise. But also that you would never probably want to see everything at once. We never really rendered it, but the idea is that you would kind of be like an explorer in it, that if you had a network which was Montréal, Milan, Zurich, Los Angeles, you would have a little loop and you would find out that you were living on membranes that connected those loops up, but you might never know that those membranes were part of these whole continents – until you started to find something else that was adjacent, and things would move around so that they would also configure in such a way that you would make your own little world within this much larger one. We all found that the image of a globe to represent the connections and adjacencies of the Internet was really off and *New City* is an attempt to find a more appropriate environment to describe the world of communication.**

Giovanna Borasi: So, in a way, the system allows you to discover yourself as part of this new geography. You might know your two membranes, but you might not know that you're part of a much more complex system. Greg Lynn: **Yeah. Even while you are visiting a place, its position changes. So, when you go back to it, it's not always the same thing you're going back to, everything is always evolving and morphing. Although maybe your little network stays familiar and consistent, there is always something out there to be explored.**

Giovanna Borasi: You assigned a project to your students related to questions about gravity and scale. How did you frame it? Greg Lynn: **There used to be a group – Neil Denari, Wes Jones and Ken Kaplan were part of it – that had a whole NASA aesthetic. Wes did the Astronaut Memorial (the Space Mirror at the Kennedy Space Center). I associate that group with the fillet. That's where the ceiling and the wall were filleted and the wall and the floor were filleted, and everything became like sections of space ships and stuff,** where it was all rounded together. I don't think that the architectural pre-occupation with filleting things came out of Rem Koolhaas' Central Library in Seattle, or Elizabeth Diller and Ricardo Scofidio's Eyebeam museum. I associate it with Neil and Wes, which was right out of this kind of NASA aesthetic.   And so my students in Vienna and UCLA – especially because Neil's here too – I could not get them to stop filleting everything together till there was no ground. And for a long time, I would say, "How do you feel about the ground relative to the ceiling or the wall? It seems like it's all the same." And I thought the space tourism problem would be a great way to get them to think about the ground: what it means to leave it, and what it means to come back to it. That's why I had them think about one project which was on a horizon, and another project which was in a dough-

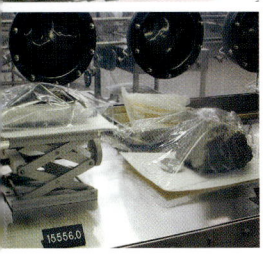

nut. And think about the difference in architecture between the station and the hotel. So it was partially about gravity, but it was actually more about ground. Like, how do you think of the ground and a horizon?

Giovanna Borasi: And were the students able to provide the answer you were looking for? Did they manage to define new ways of dealing with the conditions of gravity or no gravity? Greg Lynn: **Well, in Vienna, they only focused on the stuff in outer space. At UCLA, the outer space thing was more like a starting point, and then they all focused on the space terminal in New Mexico. And I have to say I was disappointed in the space terminals because they turned out to be like Norman Foster's thing – like a regional airport.   In Vienna I asked the students to imagine a vignette that focused on the design of the horizon. There were some pretty good ones. The one that I liked the best, you can't understand unless you see it moving: there are two rings that sit through one another, that turn through each other, and one of them is the ground. It looks like you're on the ground but then the other one rotates past you every once in a while in this totally oblique way, which makes you realize that, in fact, you're moving, and it's really great. And so there were a few little things like that that I thought were really amazing.**

case used to transport lunar
mples back to earth during the
pollo program, on display at
e National Air and Space Museum
the Smithsonian Institution in
ashington, DC.

oon rocks that have been returned
ter experiments to the Lunar
ample Laboratory Facility at the
hnson Space Center in Houston,
X. Prior to experimentation, the
cks are kept in nitrogen gas; when
amples are exposed to the earth's
umid atmosphere they react with
to form rust and clay.

Giovanna Borasi: And is this topic something that you want to continue to explore? Greg Lynn: **Yeah. I think some of the major projects right now for people to think about are horizon, ground plane, gravity, structure, vertical transportation. All those things are looser topics than they've been in a while. Much more interesting to me than facade or cladding, which is also in the air.**

Giovanna Borasi: In a recent publication, you wrote something about who is for you a visionary architect. If I remember well you were saying that Le Corbusier for example was visionary in the way he used concrete, and Mies van der Rohe in the way he used steel and glass. I think the idea of being visionary in terms of using materials or technology is interesting. Greg Lynn: **I just was looking again at Emil Kaufmann's book**

Walter M. Schirra
16834 Via de Santa Fe, Box 73
Rancho Santa Fe, CA 92067

This is my Goody comb that I used during the flight of Apollo 7 during October 11 through 22, 1968. It was part of my personal hygiene kit carried on Apollo 7. The comb was important to have while preparing for our several television broadcasts from the spacecraft. No doubt it helped me win my Emmy for those live space TV shows!

Apollo 7 was the first manned flight of the command and service modules. The mission objectives included extensive tests of all spacecraft systems and the demonstration of rendezvous capability. I was mission commander and flew with crew members Walter Cunningham and Donn Eisele.

Walter M. Schirra
Captain USN, (Ret.)
Mercury, Gemini,
and Apollo Astronaut

These are the military sunglasses that I used while flying various aircraft as an astronaut with NASA during the 1960s. The aircraft I flew during that period included the F-102, F-106, T-33, and T-38, as well as my private Beech aircraft.

Gordon Cooper
Col. USAF, (Ret.)
Mercury and Gemini Astronaut

Timed to coincide with the 40th anniversary of Apollo 13, on July 16, 2009, Bonham's auctioneers hosted "the highest-grossing American space history auction ever," featuring artifacts including flight plan sheets, emblems, lunar surface equipment, star charts and photographs, some of which had been to space and back.

on Claude-Nicolas Ledoux and Le Corbusier where he describes both architects as visionary because they were interested in Platonic geometries that couldn't be perfectly realized on earth; geometries of perfect worlds.   The other great thing about all the technologies of space exploration is the backyard invention part: these are things that can't be perfect because they have to be built. I remember when I saw the Apollo capsule, as a teenager maybe, realizing it looked like something you could build in your garage. I mean, it was really crude. All this stuff was super crude, and very – not ad hoc – but thrown together. And that's the thing I like about NASA: it's really "right thing for the right job" and not so aestheticized, and perfect, and pure, and heavenly.

In 1969, 3 astronauts travelled to the moon and back in the Apollo 10 capsule. The command module is now on display at the Science Museum, London.

Giovanna Borasi: It is true. I was looking at a publication about astronauts' objects, like their suits. They do not have a perfect look. It's what you need for that journey, and it looks like someone made it.   Greg Lynn: Yeah, it's cool. I remember seeing all the space suits. I saw an exhibition of cosmonaut and astronaut space suits and watches and all the technology they took with them, comparing the Russian and American programs. Knowing that this stuff had been in space and come back gave it this whole other, like, "Wow, I wonder if there's a little zero atmosphere left somewhere in the pocket or something?"

Giovanna Borasi: To describe an exhibition, Harald Szeemann used the term "organism," implying both the idea that making an exhibition can be a process of growth, thinking and invention, but also that an exhibition can propose new models of thinking. So, I was wondering what is your aim with this exhibition that we are doing together? Greg Lynn: Well, the first thing I thought of was... have you ever seen moon rocks?

Giovanna Borasi: No. (laughs) Greg Lynn: When they display a moon rock... it's a rock. But there's always something like a glass case that makes it look like it's in a different atmosphere. Like, the aura. Maybe it has to be that way. I don't know, maybe moon rocks aren't supposed to be in the earth's atmosphere because something could happen. I want the things in the show to seem like they're chunks of something from another place that happened to find their way into the institution. And because New City is a virtual thing, I really feel like somehow it has to look like we reached into a virtual world, and grabbed a thing, and took it out, and it's frozen or something. It should be like it's arrested, and it's not complete, it's a fragment. I want it to make you think that there's some other... something. And hopefully the videos and the technical drawings will help with that. It needs to be like, not a window into something, but like artifacts were taken out, and the minute they were taken out they froze. So, I'm hoping it has an otherworldly quality.

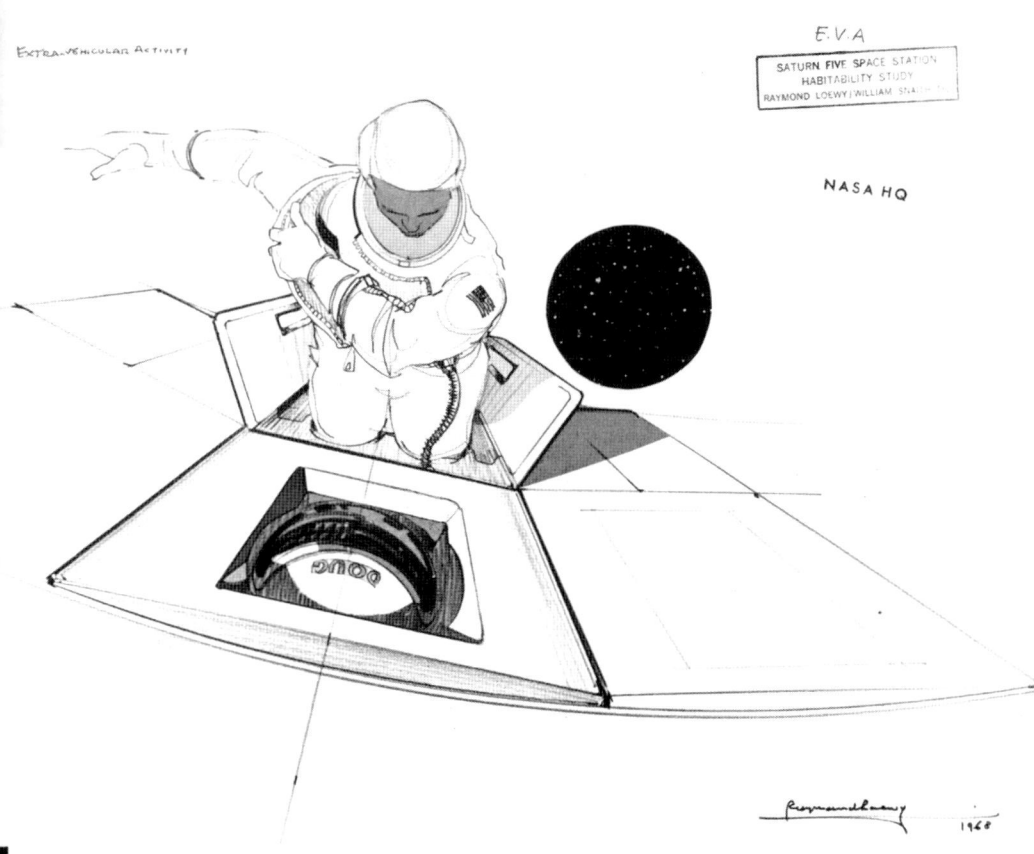

EXTRA-VEHICULAR ACTIVITY

E.V.A

SATURN FIVE SPACE STATION
HABITABILITY STUDY
RAYMOND LOEWY/WILLIAM SNAITH

NASA HQ

1968

Extra-vehicular Activity (EVA), an illustration by Raymond Loewy for *Space architecture: the work of John Frassanito & Associates for NASA/ John Zukowsky*, published in 1999 with a preface by Buzz Aldrin.

SPACE STUDIO: THE UNIVERSITY OF CALIFORNIA, LOS ANGELES AND THE UNIVERSITY OF APPLIED ARTS, VIENNA

Greg Lynn

I recently led classes – at the Department of Architecture and Urban Design at UCLA (University of California, Los Angeles) and at the University of Applied Arts in Vienna – exploring the design of space colonies. The students and I imagined a research project that started with a historical overview – from Wernher von Braun and Gerard O'Neill's visionary proposals, to NASA's plans for a space station on the moon and the implications of space tourism. Next we focused on the challenges involved in meeting the current conditions presented by artificial space habitats, such as synthetic gravity, levels of oxygen pressure, transportation, etc.

Besides architectural design, the assignments addressed systems for sustainable living; alternative life styles; and political and cultural issues in an entirely man-made environment, potentially unencumbered by the typical layers of tradition that characterize life on earth. Implicit in the challenge to develop new societies from the ground up is the opportunity to rethink the discourse about the current terrestrial condition, which might ultimately benefit from the design of these extraterrestrial prototypes.

Sections and plans of space colonies designed
by Greg Lynn's students at the University of Applied
Arts Vienna, 2008.
Right: Lisa Sommerhuber
Page 58: Elisabeth Brauner
Page 59: Adam Lind

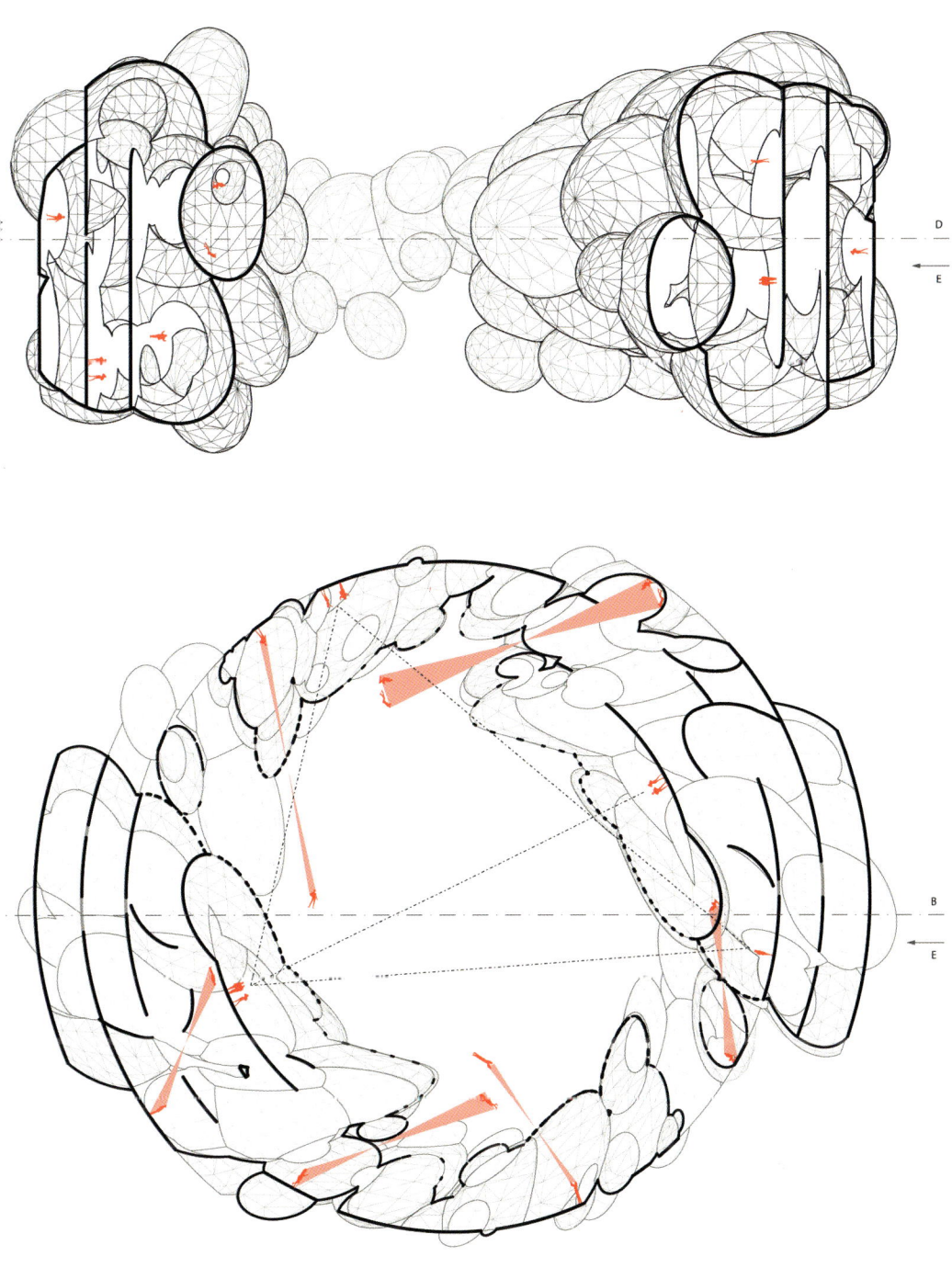

D
←
E

B
←
E

57

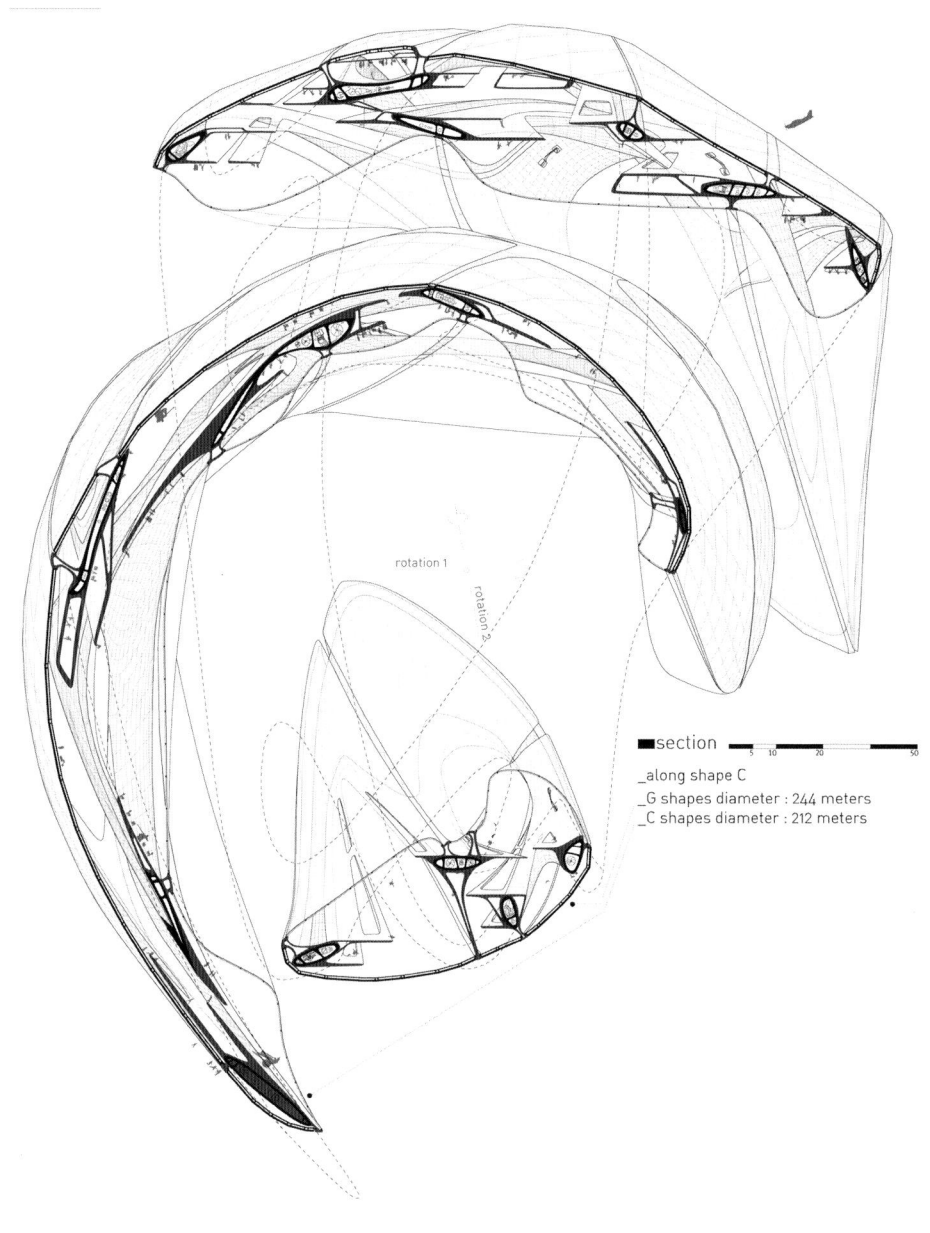

rotation 1

rotation 2

■section   ┣━━┿━━┿━━┿━━━┫
          5 10   20   50

_along shape C
_G shapes diameter : 244 meters
_C shapes diameter : 212 meters

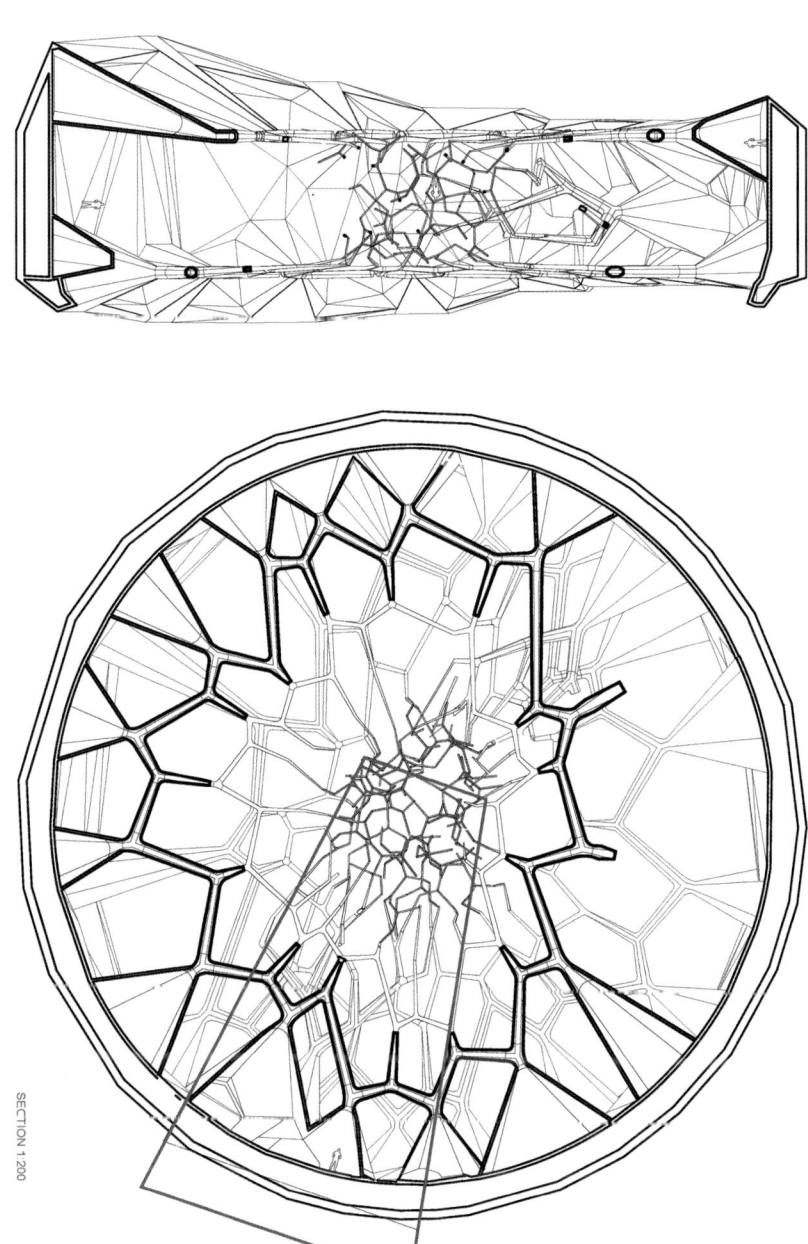

SECTION 1:200

ZENO'S SHED IN OF RIPARBELLA R SPACE CAPSULE.

ALESSANDRO POLI

E COUNTRYSIDE
EMBLED ·ALDRIN'S

ALESSANDRO POLI (born in 1941 in Florence) lives and works as an architect in Florence. He holds a Master of Architecture degree from the Faculty of Architecture at the University of Florence.

From 1970 to 1972 he was part of the Italian radical architecture group Superstudio (Adolfo Natalini, Cristiano Toraldo di Francia, Roberto Magris, Gian Piero Frassinelli, Alessandro Magris).

From 1973 to 1982 he taught at the Faculty of Architecture in Florence, focusing in 1974–80 on the research for *Cultura materiale extraurbana*. Part of this work was presented at the 38th Venice Biennale (1978) with Superstudio.

His current research investigates the relationship between art and architecture at different scales: lately he has been focusing on the design of exhibitions and of contemporary art jewelry.

1 alluna aggio

La luna è
uscita dalla
sua sfera.
verginale

Storyboard by Alessandro Poli for the movie
Architettura interplanetaria, Superstudio, 1970–71
Birth of the earth and moon and subsequent
phases of adjustment

Right: Hypothesis of enlarging the terrestrial
surface to restore the earth's original integrity
Handwritten storyboard: pencil and ink on
notebook

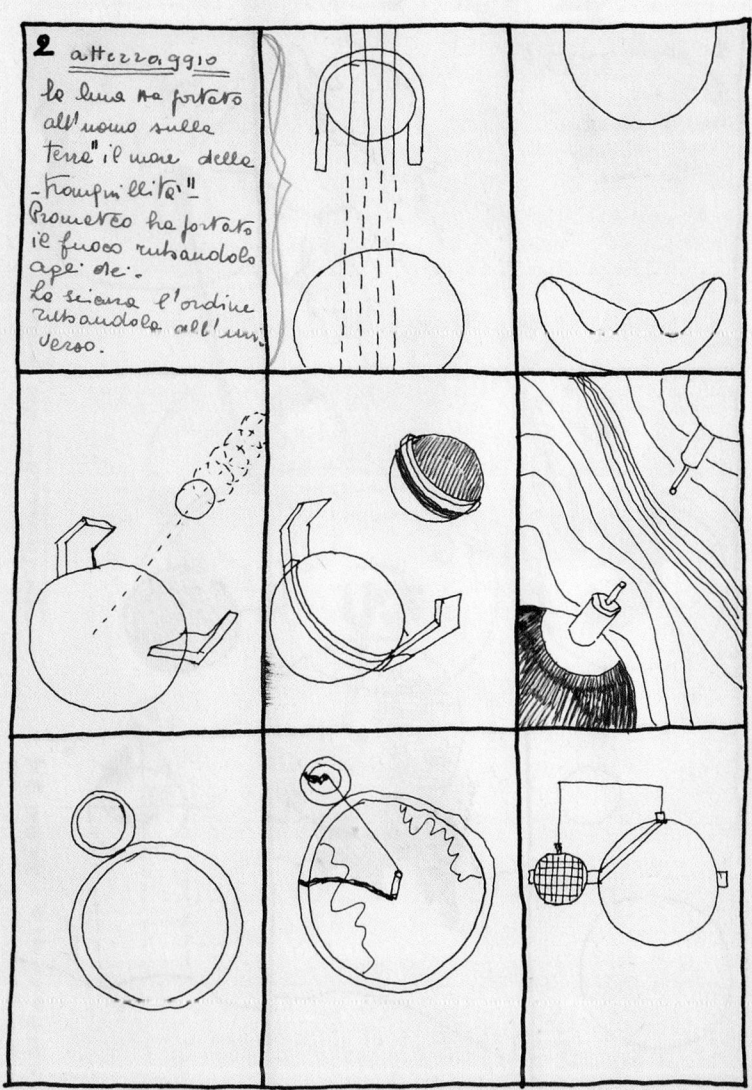

SUPERSTUDIO PRESENTS

L'ARCHITETTURA INTERPLANETARIA

(Originally published in Casabella, *n. 364. April, 1972, p. 46–48)*

1. Birth of the earth and the moon and the successive settling (first observation for a hypothesis of enlargement of the terrestrial surface)
4 billion years ago the day was 4 hours long and the moon was in an orbit with the radious of approximately 14,000 km. In as much time again if man maintains his ecological planetary system intact, the solar day and lunar month will be of the same length, that is, equal to 60 days of present time. Thus the day will be 1,440 hours long. There are hypotheses of the birth of the moon from the earth as the separation of a mass of material from the earth. The moon then gradually withdrew until it reached its present orbit.

2. Observation by man of the earth-moon system. Man's arrival on the moon. The first footprint on lunar soil. Cognitive possession (second phase for a hypothesis of enlargement of the terrestrial)
The reader is referred to the extensive NASA chronicles for details and to all the pictures diffused by widely read channels. Contribution to the event may be made with some proposals for lunar improvement, improvement, that is, of the public image of the moon.

3. Nearing the moon to the earth; their coupling (third phase for a hypothesis of enlargement of the terrestrial surface and for the restitution of its natural virginity to the earth)
The moon is brought to a distance of approximately 40,000 km, equal to its diameter. At this distance the moon is relatively fixed at (facing) the same meridian and parallel and the eclipses of the sun will be more frequent. The moon will revolve around the earth in one day. At this distance from each other, earth and moon are considered a single body.* The minimum distance possible between moon and earth is 16,000 km. At less than this, the action of the tides would destroy the moon. The distance of 16,000 km is calculated on Roche's limit, which declares that the minimum distance is equal to the cubic root of six times the earth's mass divided by pi times the lunar density.

4. Earth-moon highway with capture of wandering meteorites to create the gravitational axis fixed around the highway strip
The earth and the moon, at a distance of 40,000 km, form a geostationary system: their distance is, in fact, the same as that of those satellites which, since they have the same period as the earth, we see fixed on the horizon. Two artificial belts around the earth and the moon respectively are coupled together permanently with a rectilinear body.

5. Enlargement of the terrestrial surface with the capture of small wandering planets drawn into the Earth's orbit and relative coupling with the Earth

The group of small planets to be used is found in a strip compressed between Jupiter's orbit and that of Mars. These small planets, called asteroid, are thought to be the remains of either a planet that was never formed or one disintegrated after formation by the passage of a comet. This can be deduced by the law of the Titius Bode (1972) which says that the distance of a body (planet) from the sun in our total system is: $d = 0.4 + 0.3 \times 2^n$. When for n we substitute minus infinity for Mercury, 0 for Venus, 1 for the earth, 2 for Mars, the strip of asteroids (observed by Patti in 1801) appears in the third place.

*See the mathematical formula below.

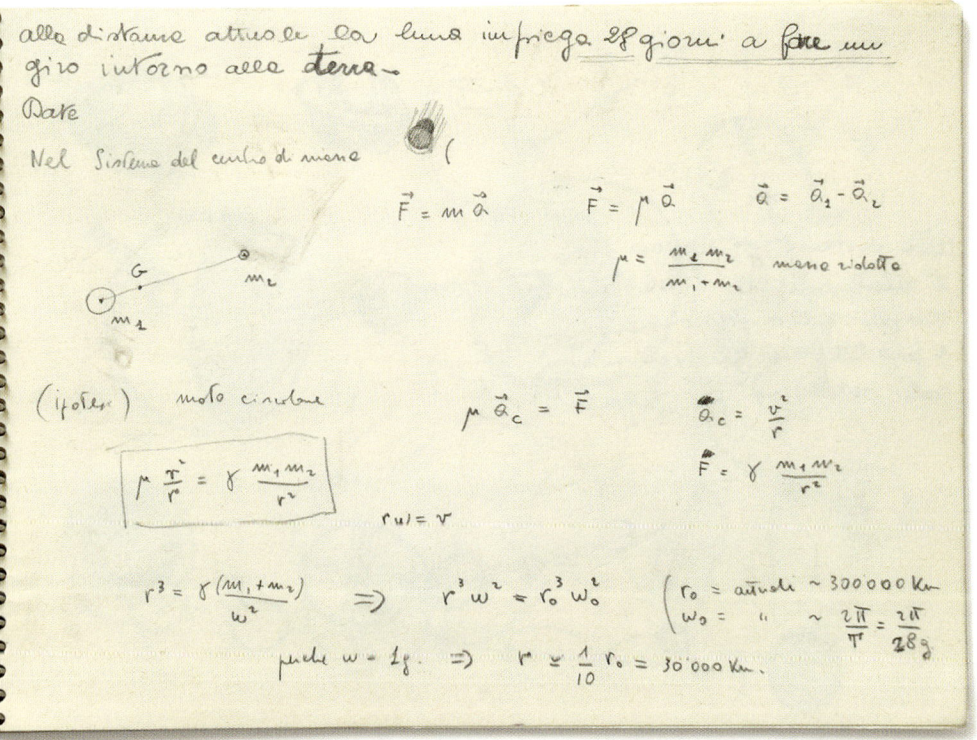

Notes by Alessandro Poli for the movie
Architettura interplanetaria, Superstudio, 1970–71
Utilizing the universe and its energy for a global
architecture

Handwritten calculations: pencil and ink on
notebook

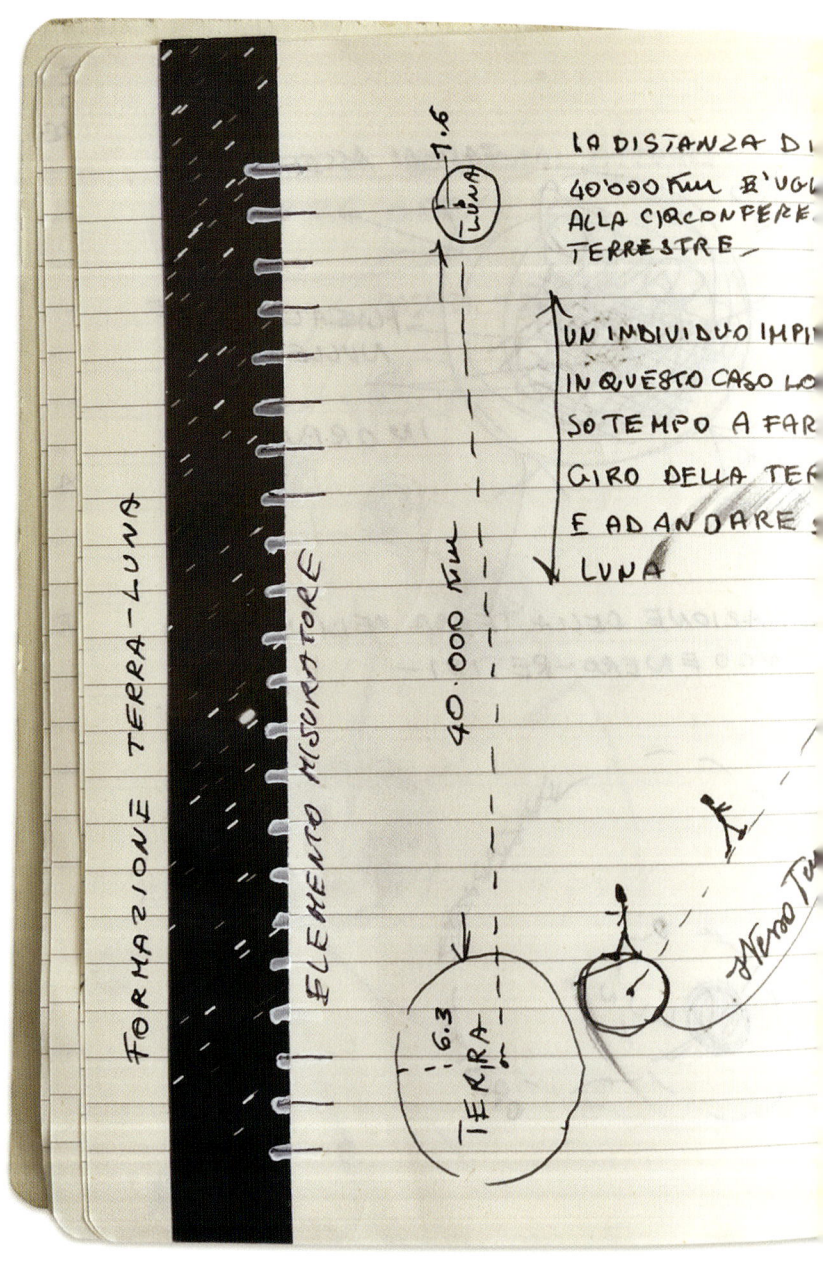

Notes by Alessandro Poli for the movie
Architettura interplanetaria, Superstudio, 1970–71
Hypothesis of nearing the moon to the earth

The moon is positioned 40,000 km from
the earth, a distance equal to its circumference.
Ink sketches in ruled exercise book

ARK COME USO-ENERGETICO-
L'ENERGIA E' IL PUNTO CENTRALE
DELLA NUOVA ARCHITETTURA-

L'USO DELL'ENERGIA E LA SUA
DISTRIBUZIONE E LA NUOVA
ARK. CHE USA IL COSMO COME
RIFERIMENTO-

L'ARK INTERPLANETARIA NON E'
UN'ARCHITETTURA COSTRUITA
CON LE REGOLE CONOSCIUTE...

MA E' L'USO DEL COSMO E
DELLA SUA ENERGIA COME
ARK. GLOBALE_

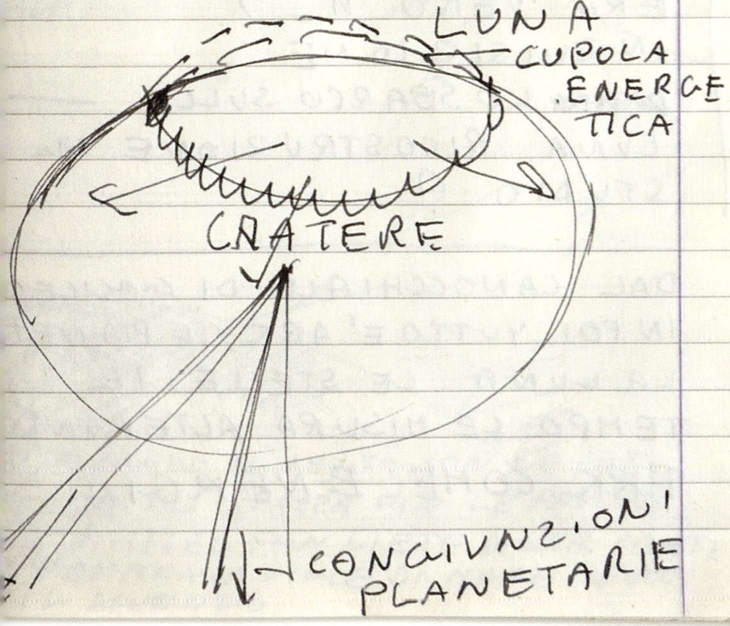

LUNA
CUPOLA
ENERGE
TICA

CRATERE

CONGIUNZION'
PLANETARIE

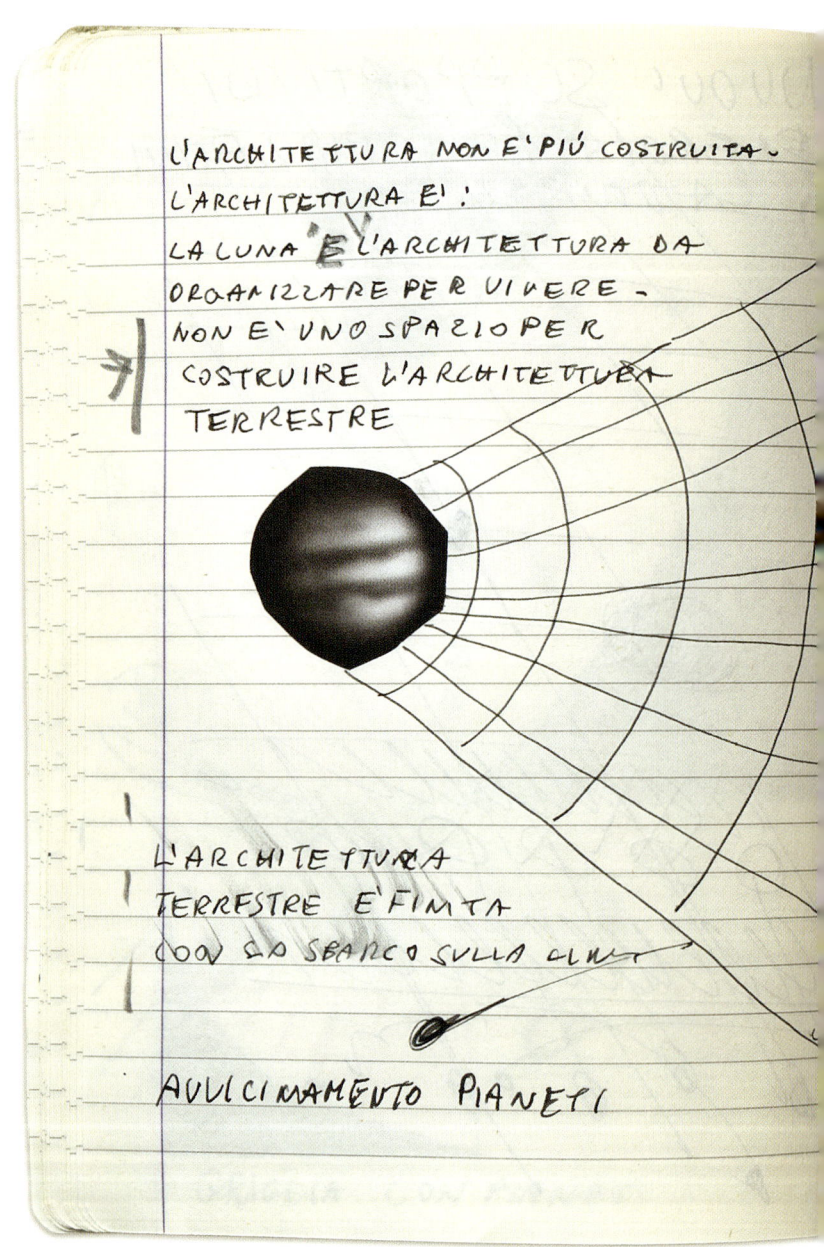

L'ARCHITETTURA NON E' PIÚ COSTRUITA.
L'ARCHITETTURA E' :
LA LUNA E L'ARCHITETTURA DA
ORGANIZZARE PER VIVERE .
NON E' UNO SPAZIO PER
COSTRUIRE L'ARCHITETTURA
TERRESTRE

L'ARCHITETTURA
TERRESTRE E' FINITA
CON LO SBARCO SULLA LUNA

AVVICINAMENTO PIANETI

Notes by Alessandro Poli for the movie *Architettura
interplanetaria*, Superstudio, 1970–71
The earth and moon when only 40,000 km apart
are conceived of as one element.
Ink sketches in ruled exercise book

Notes by Alessandro Poli for the movie *Architettura interplanetaria*, Superstudio, 1970–71 Many drawings and photomontages were kept as study notes as they were not used for the film. *Right:* Lunar landscape with view of the Sea of Tranquility. *Pages 72–75:* Everything is architecture: the moon, the planets and the stars Photomontages, sketches, and literary narrative on an insert of an illustrated Italian weekly.

Pages 74–75: Avvicinamento dei pianeti e fasi di assestamento (Nearing the earth to the moon and phases of adjustment), Superstudio (Adolfo Natalini, Cristiano Toraldo di Francia, Roberto Magris, Gian Piero Frassinelli, Alessandro Magris, Alessandro Poli [1970–72]), *Architettura interplanetaria*, 1970–71 Photomontage

Pages 76–77: Autostrada Terra-Luna (Earth-moon highway), Superstudio, *Architettura interplanetaria*, 1970–71 Photomontage

Page 78: Autostrada con sinusoide di energia (Earth-moon highway with sinusoid of energy), Superstudio, *Architettura interplanetaria*, 1970–71 Photomontage

Page 79: Trasporto dei pianeti per ingrandimento della Terra (Moving the planets for the enlargement of the terrestrial surface), Alessandro Poli, 1973 Photomontage

Pages 80–81: Autostrada – avvicinamento pianeti (Highway nearing the moon to the earth), Superstudio, *Architettura interplanetaria*, 1970–71 Photomontage

Page 82: Autostrada di collegamento fra asteroidi (Highway connecting asteroids to each other), Superstudio, *Architettura interplanetaria*, 1970–71 Small planets are carried from the orbits of Mars and Jupiter Photomontage

Page 83: Architettura riflessa (Reflected architecture), Superstudio, *Architettura interplanetaria*, 1970–71 Photomontage

Page 86 top: Nuove architetture con avvicinamento di masse planetarie (New architectures with nearing of planetary masses), Superstudio, *Architettura interplanetaria*, 1970–71 Photomontage

Page 86 bottom: Costruzione di un edificio gonfiabile (Construction of an inflatable building), Superstudio, *Architettura interplanetaria*, 1970–71 Photomontage

Page 87 top: Nuove architetture lunari (New lunar architectures), Superstudio, *Architettura interplanetaria*, 1970–71 Photomontage

Page 87 bottom: Paesaggio lunare – Luna Park (Lunar landscape – Luna Park), Alessandro Poli, 1973 Photomontage

Pages 88–89: Nuovi paesaggi rurali (New rural landscapes), Superstudio, *Architettura interplanetaria,* 1970–71 Photomontage

ALLUNAGGIO — MARE DELLA TRANQVILLITA — (CIELI ROVESCI) MANUALE DI BORDO
① INVTAZIONE ↓

STORY BOREA MANUALI → DI BORDO — TUTTO E' ARCHITETTURA — IL PIANETA — LA LUNA — LE STELLE IL TEMPO LE MISURA MVTA NOCCE

ALDRIN DOPO AVER PIANTATO SULSUOLO LA BANDIERA DEGLI STATI TI ARRIVA AL MARE DELLA TRANQUILLITAI — L'ARCHITETTURA E' IMMOBILE — IL CIELO RIFLESSO E SOSPESO — IMMAGINE #4 —

FILM - ARK - INTERPLANETARIA - ESAME DEI MATERIALI - TUTTO E' FUORI SCALA - GRANDE SCHERMO   TUTTO E' ARCHITETTURA - LA LUNA - LE STELLE I PIANETI SO...

Dal distacco dalla capsula Columbia al momento i...

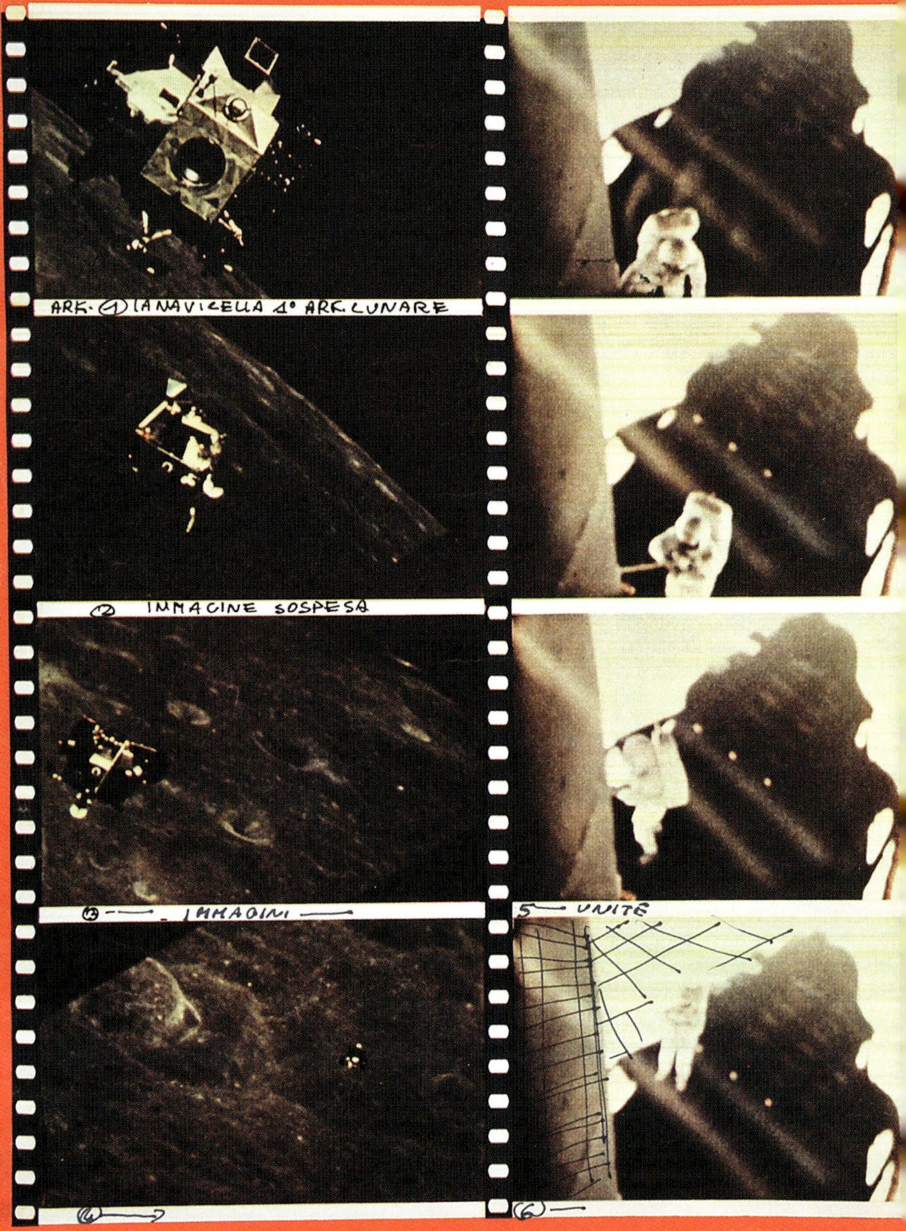

ARK. ① LA NAVICELLA 1° ARK LUNARE

② IMMAGINE SOSPESA

③ - IMMAGINI -

⑤ - UNITÉ

⑥

La navicella del LM si è staccata dalla capsula e si avvia verso la superficie della Luna. A poco a poco l'Aquila diventa più piccola e infine quasi scompare. La grande avventura è alla sua fase conclusiva. Nella seconda colonna: la telecamera fissata sul LM riprende lo storico istante in cui il comandante Armstrong esce dall'Aquila per affrontare l'esplorazione lunare.

GINE E' UN PEZZO DI ARCHITETTURA PERCHE' ORA'ORA'CHE TUTTO SI APPARSO SU
DI ARCHITETTURA

I lascia il suolo lunare, terminata la sua impresa

MONUMENTO ALLA RESISTENZA

ARCHITETTURA RIFLESSA

APPARIZIONI

I due astronauti, Armstrong e Aldrin, hanno piantato sul suolo lunare la bandiera degli Stati Uniti. Si apprestano a compiere il lavoro assegnato. Nella seconda colonna: il LM si è già staccato dalla superficie del nostro satellite per affrontare il viaggio del ritorno. Sulla Luna rimangono gli apparecchi e la bandiera. Il segno che l'uomo ha, per la prima volta, raggiunto un corpo celeste.

73

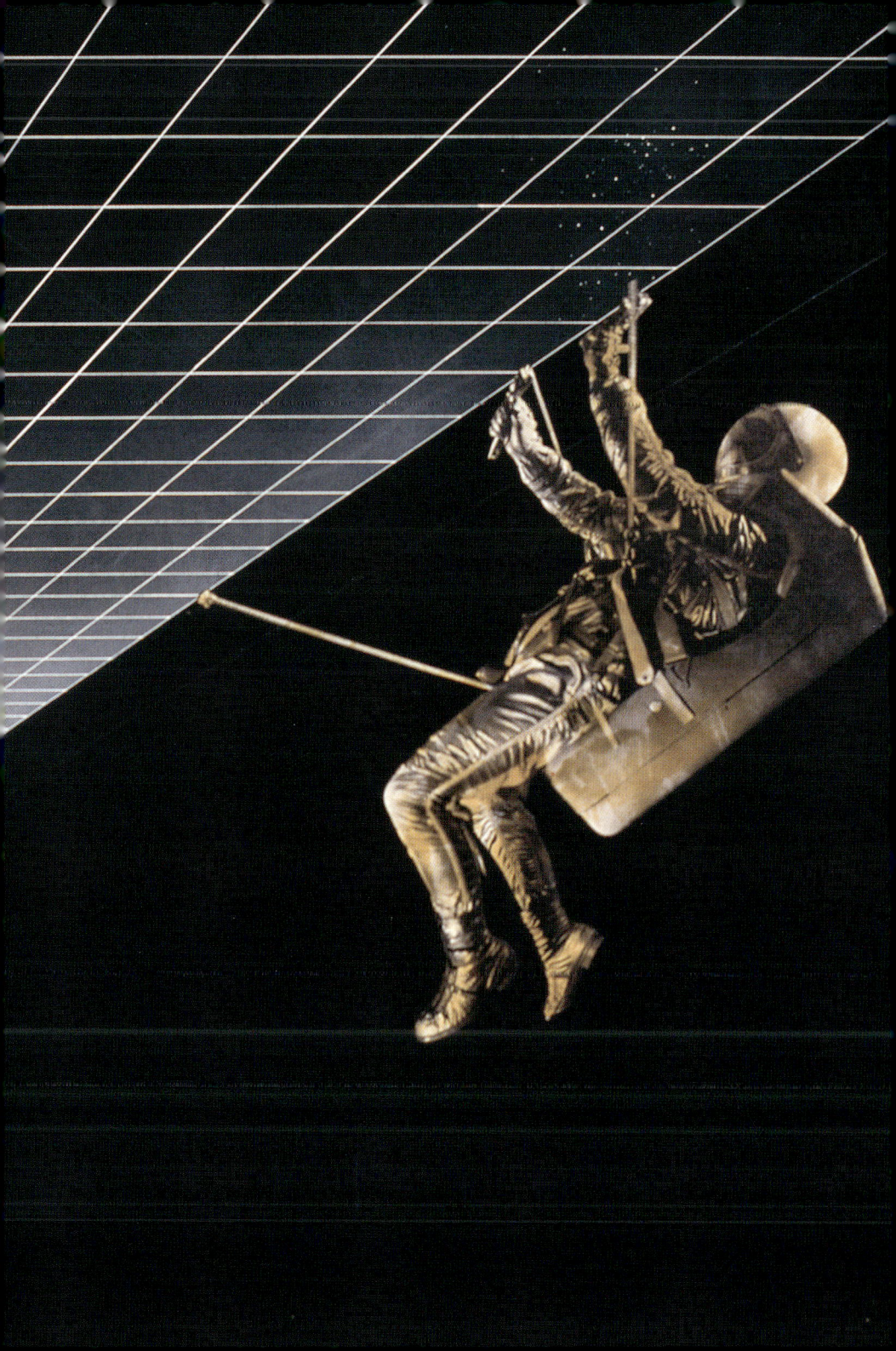

Florence, 1970

Dear Adolfo,
We saw the conquest of space on TV, the
greatest media event of the twentieth century.
After the landing on the moon, architecture
can no longer be as we had thought-imagined-
constructed it in our visions: from now on, the
real image surpasses the fantasies, our utopias,
which guided our creations.

No clouds, no wind, no gravity, no conflicts,
no sounds, and yet the tremendous strain
of discovering that we are small because even
huge monuments, solid architecture, seem
so far away that they vanish.
Rather than conquering space, it is space that
has conquered us, leaving in us a swath of
unfinished projects, memories, designs never
executed, possibilities, the difficult return to
earth and life.
The architecture of the spaceship is made from
nothing, covered with tinfoil, as fragile as
its occupants, yet extremely fascinating in that
dusty landscape, where everything seems
motionless.

Interplanetary architecture is landscape
architecture: everything is architecture, the
planet, the moon, the stars.
Our project will be to compose, to understand,
to let ourselves be carried away, to travel
by new means and to reimagine surroundings
to live in, not with the anxiety of creating new
architectures and monuments, but as huts
where we can live engulfed in the architecture
of space.

To design this, we will need different media.
No more pen and ink, pencil or photomontage:
we will need the motion of the movie camera.

Firenze '70

Caro Adolfo

abbimo visto in TV la conquista dello spazio,il più grande evento media
tico del '900.

Dopo lo sbarco sulla Luna il progetto d'architettura non può più
essere come lo abbiamo pensato-immaginato-costruito con le nostre
visioni; l'immagine reale ormai supera quella fantastica con laquale
abbiamo progettato i nostri sogni; le nostre utopie

niente nuvole,niente vento,niente forza di gravità,nienteconflitti,
niente rumori, eppure grande tensione di scoprirsi piccoli,deve-aneh
perché anche i grandi monumenti ,le solide architetturee-nelle
quliei annullano fino a sembrarci lontani.
altro che conquista dello spazioda-p, perché é lo spazio che ci ha e
conquistati lasciando in noi una scia di progetti incompiuti
di memorie , dip progetti mai realizzati, di possibilità, di diffi-
cil-e ritorno a terra alla vita;
L-- L'architettura dell'astronave é fatta di niente rivestita
con carta stagnola, fragile come fragili erano i suoi abitanti
eppure sestremamentes--affascinante in quel paesaggio polveroso
dove tutto sembra immobile.

L,architettura interplanetaria é architettura di paesaggio:
dalla-eem-tutto é architettura il pianeta la luna le stelle

iln il nostro progetto sarà quello di comporre capire andare lasciandosi tra
trasportare e attraversare con mezzi e ricreare ambienti dove vivere
senza ancoscia di creare nuove architetture e monumentima capanne dove
vivere immersi nell'architettura dello spazio;

Per progettare questo c'é bisogno di mezzi diversi ei-che non
sono più solo le penne a china, le matite, e i fotomontaggi,
ma c'é bisogno del movimento della maccina da presa del cinema.

MARCUSE

Notes for a letter to Adolfo Natalini, drafted
by Alessandro Poli in 1970, and never sent.
Typewritten document

Autoritratto con riflessa autostrada Terra-Luna
(Self-portrait with reflection of Earth-moon high-
way), Alessandro Poli, 1973
Photomontage

Zeno in his house in Riparbella, Alessandro Poli,
Zeno: Research on a Self-Sufficient Culture,
1979–80

ZENO AND ALDRIN MEET IN RIPARBELLA, TUSCANY, IN THE SUMMER OF 2008

Zeno and Aldrin meet, many years after the moon landing. The landscape of Riparbella is reconstructed by their conversation as they speak. The countryside of Le Preselle – with the shed, spring and row of vines – now belongs to history. The accelerated pace of contemporary times has transformed it into a nondescript suburb.

Z.   I don't recognize anything, and yet there is still much that belongs to me.
A.   Could you start again, and transform what seems to have been destroyed, using your repurposing techniques?
Z.   Possibly, but only if you succeed in altering your invasive, dominating technologies and leave me room for my work. I can teach you. You just have to believe!

Their meeting transforms the anonymous suburb where they stand into an actual place, which does not mean returning to the past but rather putting the current situation to use according to different rules, dictated by living together and understanding cultures and technologies that originate from different realities and backgrounds.

A.   I played a leading role in the greatest technological feat of the twentieth century. We set foot on the moon.
Z.   How well I remember! I thought of you while I was shut up in the shed at Le Preselle, and I feared for you, for your safe return.
A.   Our technology was safe. It had been put through every possible test, and everything functioned.
Z.   One essential factor was missing: "necessity," which is the motive underlying all our work. My deep perception of the moon, like everything I do, is rooted in necessity. This is the main motive prompting me to act, to experience, to strive according to my potential.

Zeno is the principal author of the reappropriation that could return the region to a dimension in which everything, including architecture, arises from necessity, from doubt, from respect for simple techniques even while increasingly sophisticated and invisible technologies are used. The planet has already been invaded by a media network that transforms it from real to virtual.

Z.   The greatest challenge for all is the return to earth, and we must do it together.

A.   To begin again from necessity in order to restore the possibility of redemption, of transformation through techniques of reuse, for both the artifacts of the squalid suburbs and the super-architecture that is invading the deserts.

Z.   I can guide you along the shortcuts I built to return home by in the countryside of Riparbella. They were necessary, like the moon that lit my way.

The concept of landscape borne of their meeting is a fusion of different worlds that expands space rather than compressing or invading it: it is the initial situation that arises from a conversation, from an exchange of experiences, before becoming a reality.

Avvicinamento dei pianeti (Nearing the moon to the earth) in our photomontages was an idea for merging different worlds to create alternative possibilities more than specific territories.

Alessandro Poli
Florence, January 28, 2010

Pages 94–95: Zeno incontra Aldrin a Riparbella
(Zeno and Aldrin meet in Riparbella)
Alessandro Poli, 2008
Photomontage and sketch with pencil

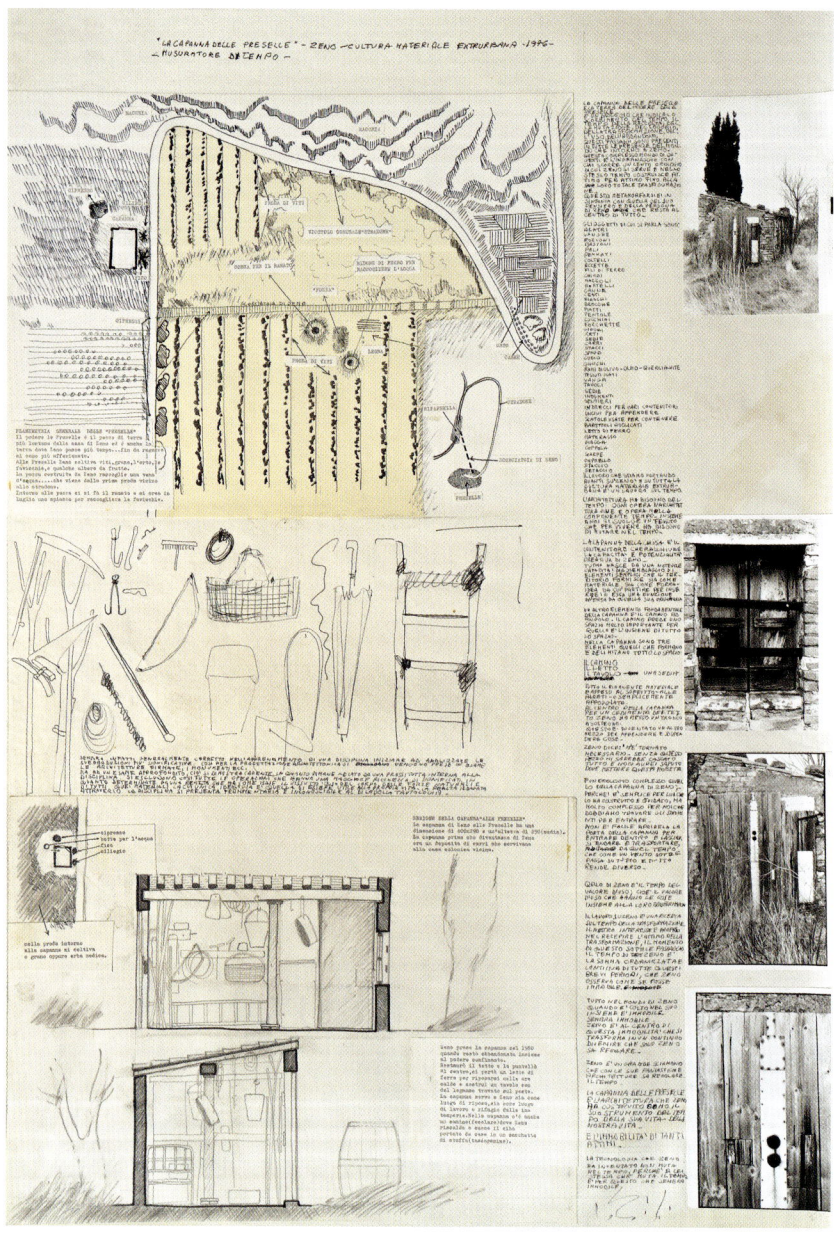

The shed and farm at Le Preselle, Alessandro Poli, *Zeno-research of a self-sufficient culture*, 1979–80
A clock that tells the time and seasons, which only Zeno is able to use and transform.

Mixed media: pencil drawing, photos and typewritten text

The shed at Le Preselle, Alessandro Poli,
Zeno-research of a self-sufficient culture, 1979–80
Detail of the door
Mixed media: pencil drawing and writing, photo,
metal piece of original door

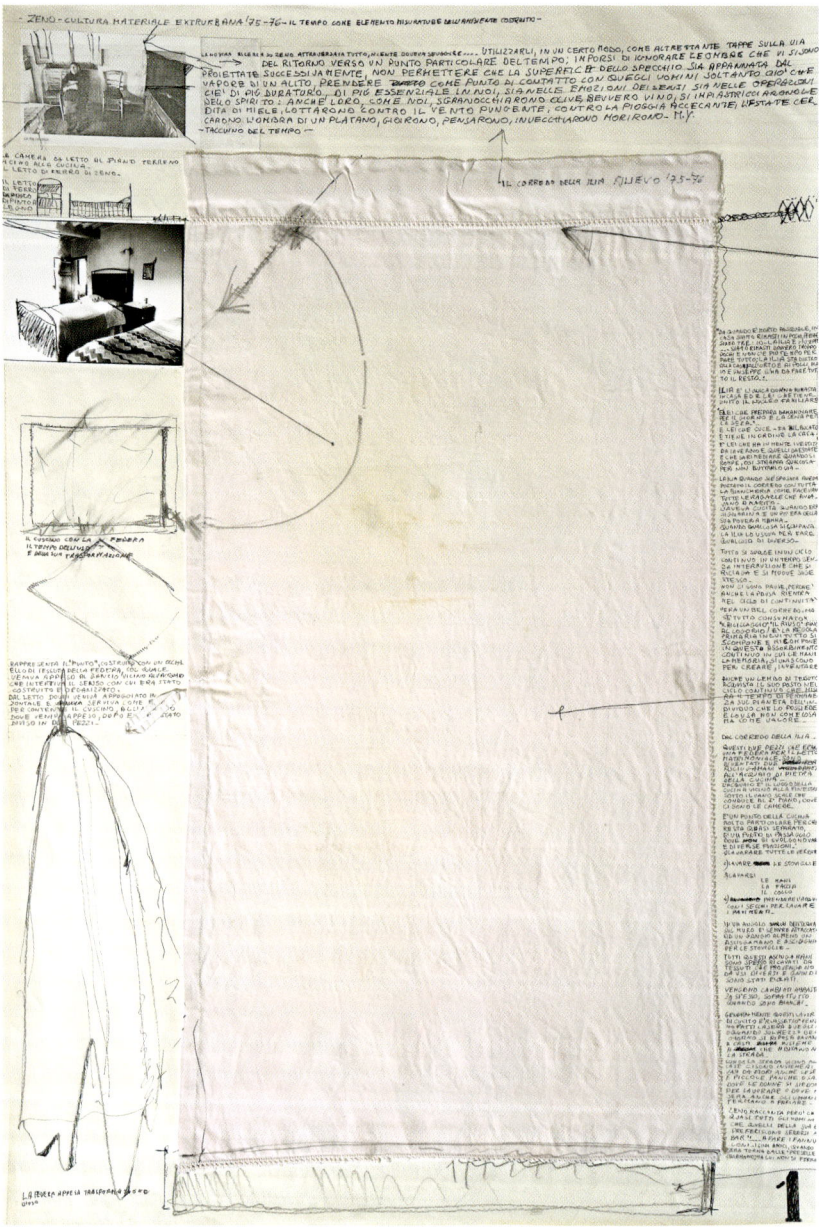

Re-use technique, Alessandro Poli, *Zeno-research of a self-sufficient culture*, 1979–80
Towels made out of reused linen from old pillows, woven on the house looms.

Mixed media: pencil drawing and writing, photos, linen cloth artifact

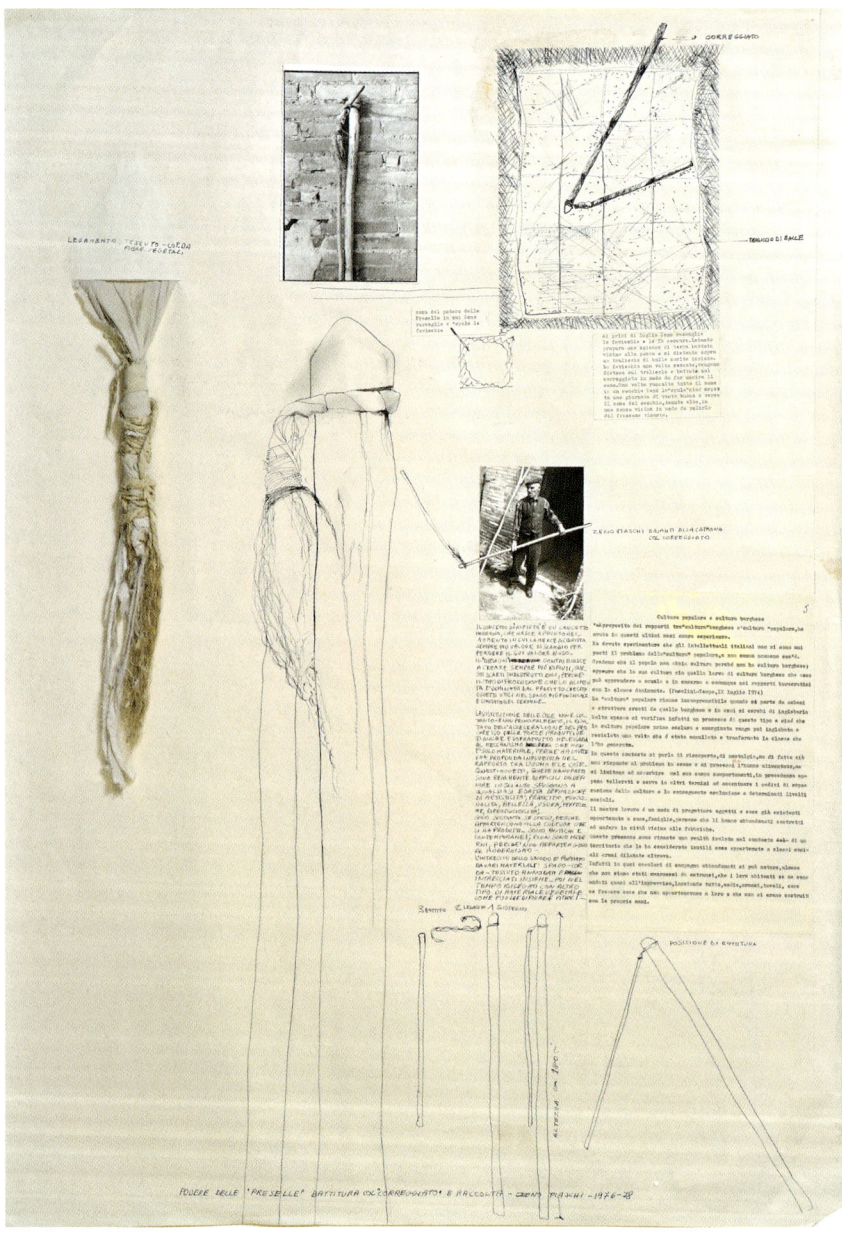

Harvesting tools assembled from various parts
and materials, Alessandro Poli, *Zeno-research of
a self-sufficient culture*, 1979–80
Mixed media: pencil drawing and writing, photos,
typewritten text and various materials

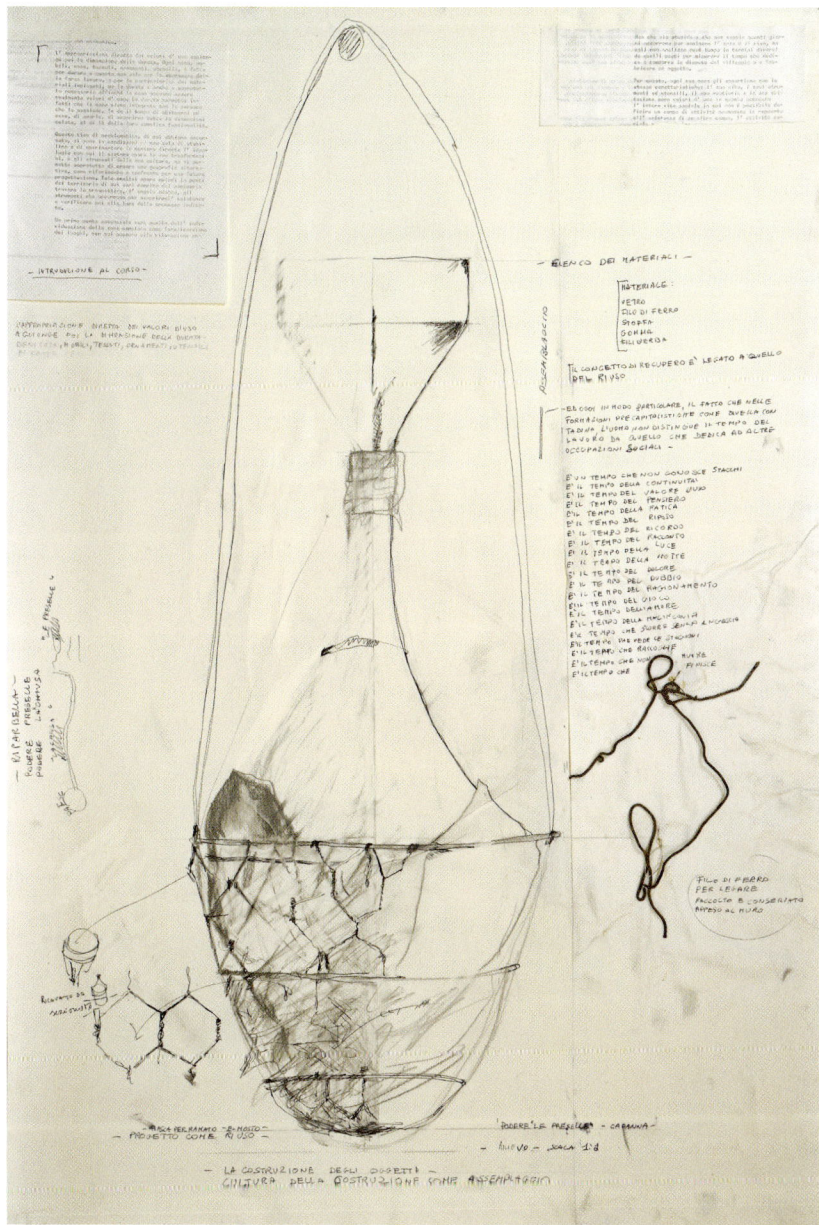

Study of a container for liquids Alessandro Poli, *Zeno-research of a self-sufficient culture*, 1979–80 Mixed media: pencil and ink drawing and writing typewritten text and piece of wire

Pages 102–103: Pieces to be reused, Alessandro Poli; *Zeno-research of a self-sufficient culture*, 1979–80 These were hanging in Zeno's workshop. This is an "illustrated catalogue" of a culture that only its author could have built, organized, reused and consumed.

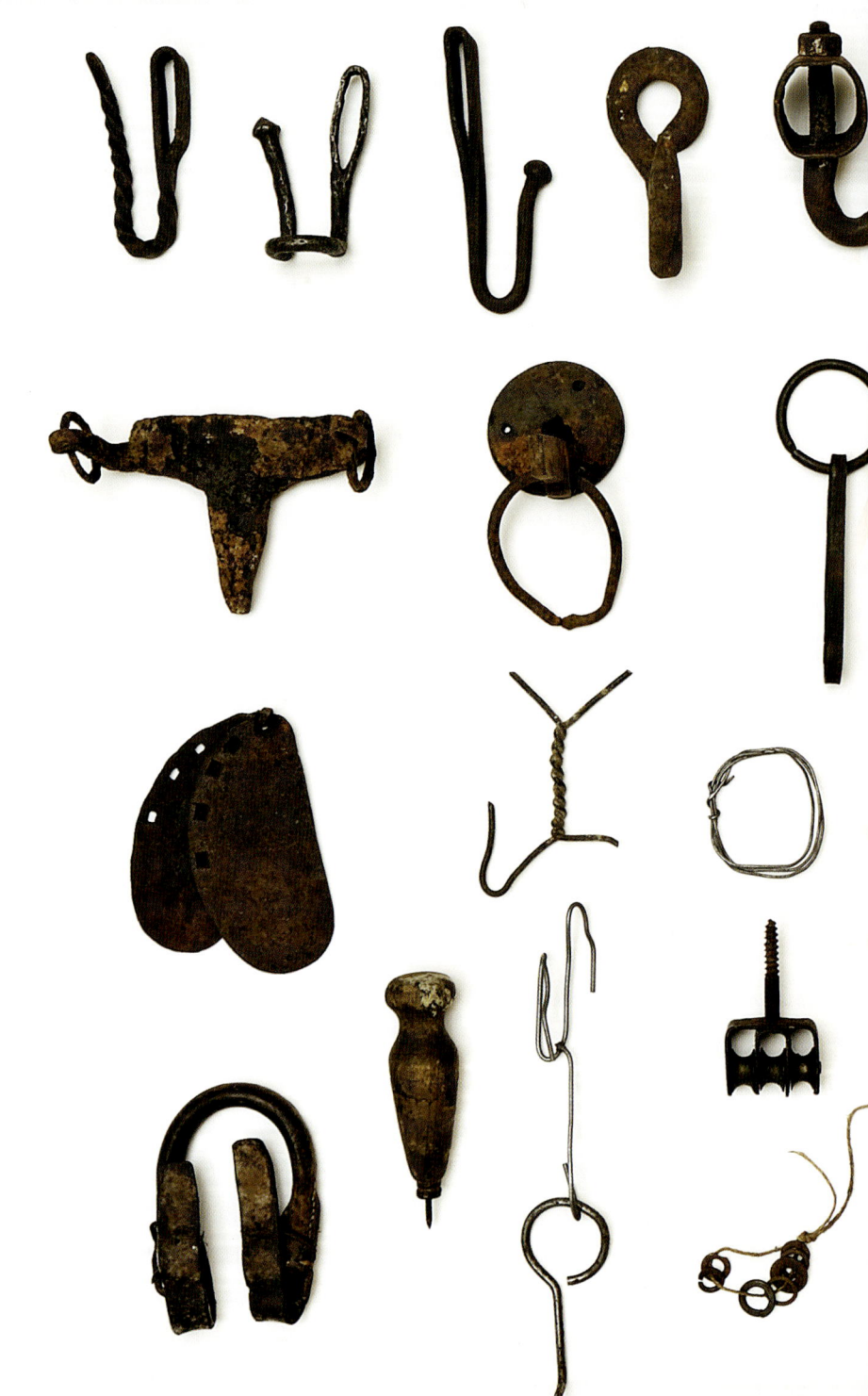

Various parts assembled to create a tool for retrieving
objects that have fallen into a well. Alessandro Poli,
Zeno-research of a self-sufficient culture, 1979–80

Container for preparing the copper sulphate solution used to treat the vineyard. Alessandro Poli, *Zeno-research of a self-sufficient culture*, 1975–80

This object was found in the shed at La Chiusa in 1980. Composed of different elements, it represents the importance of use value in Zeno's culture.

Re-use technique, Alessandro Poli, *Zeno-research of a self-sufficient culture*, 1979–80.
Left top: Various materials used to create weavings for rows of vines and to separate different crops.

Left, bottom: Strainer with cloth for filtering powders and liquids to treat plants.
Above: Funnels assembled for creating infusions of herbs and roots.

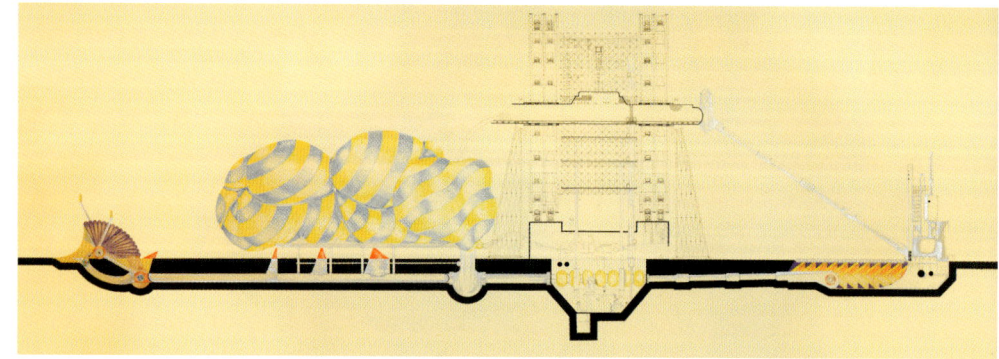

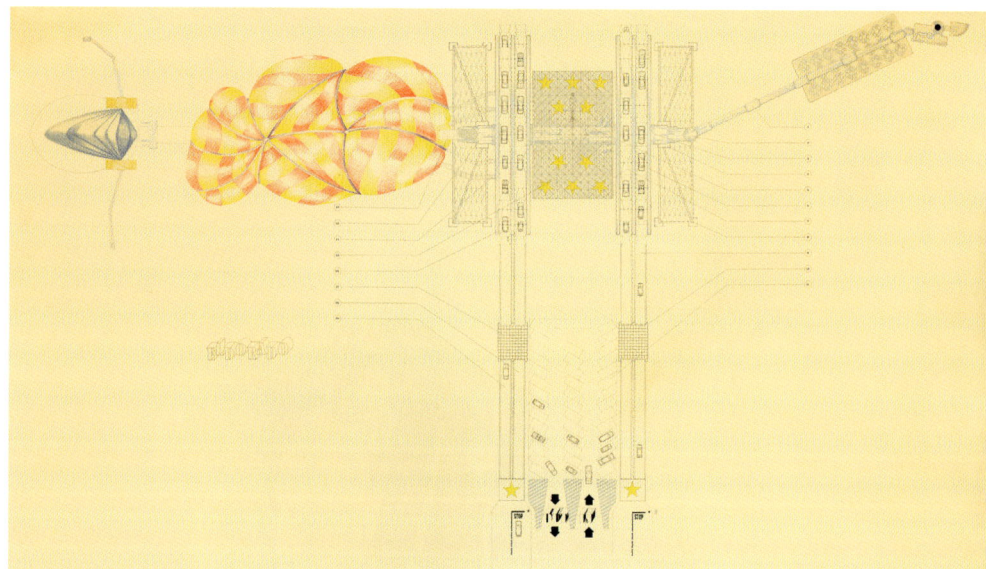

Design of a building for leisure time, R Gherardi,
S Pacini, A Poli, R Russo, F Spinelli, *Piper*, 1966
Longitudinal section and plan
Heliography on card, coloured pencils

NEARING THE MOON TO THE EARTH

Alessandro Poli

I remember as if it were yesterday hearing about the lecture Herbert Marcuse gave at the Free University of Berlin in 1967. He had announced the end of utopia – not its ultimate defeat but its impending realization. Not long before, I had finished an assignment with my group[1] as part of research being conducted by Professor Leonardo Savioli at the Faculty of Architecture of the University of Florence. The subject was leisure time, focusing on the study of a recreation and entertainment venue called *Piper*.

This was a moment when architecture was becoming aware of all the available possibilities, including new technologies and the ability to apply them in an original discourse, free of formal and functional constraints, but also apt to invent new models of behaviour wherein both machines and everyday objects could be used in an extraordinary creative way.

We began concocting coloured architectures and out-of-scale objects for a world on the threshold of magic years full of dreams, great passions and change, where architecture would play a fundamental role. Our pop[2] project belonged not to realism but to the invention of alternative living places for a world in which, as announced in Berlin, utopia was no longer a fantastic dream. Marcuse believed "that we can now speak of utopia only … when a project for social change contradicts real laws of nature. Only such a project is utopian in the strict sense, that is, beyond history – but even this 'ahistoricity' has a historical limit."[3]

This meant the confirmation of our research and projects as a possible path, rich in ferment and ideas, to be built with others (such as Archizoom and UFO) who, like us, were experimenting with new models, often derived from different cultures – the beat, hippie, underground and pop movements – that were expressing a refusal to participate passively in the benefits of affluent society, because they were seeking a qualitative change of need.

The Piper project was the large architecture-machine with which I initiated my voyage into territories considered off earth's official routes, but that others were already travelling, or had travelled and were forced to abandon because they ran counter to the prevailing ideology.

I met my friends from Superstudio and began to collaborate and compare irony- and paradox-based strategies with them, designing lamp-architectures, machine-architectures, architecture as false objects, writing and drawing architecture.

I got to know Archizoom, UFO, and 9999 and saw the images of Archigram's *Instant City* and *Plug-In City*. I also rediscovered the forgotten world of Russian constructivism – Vladimir Tatlin, Alexander Rodchenko and El Lissitzky, who, like us, had attempted to merge art, technology and architecture.

The Piper project was drawn in India ink, according to the classical techniques and functional rules of representation, on 100-by-200-centimetre sheets of tracing paper heliographed onto card and then coloured in minute detail with coloured pencil, where broad skies with clouds and rainbows often appeared along with the girders and bearing structures. It was this expressive use of colour that deformed the aseptic traditional black-and-white languages of architecture to the point of making them look like childlike oddities. To the objection often addressed to us even by our teachers – "But this is not architecture!" – it was easy to respond, "Why not?" I have always thought this ambiguity was the reason our work was so original and stimulating.

Such a utopia also shaped my thoughts and characterized my work with Superstudio, where together we sought new strategies for moving past the reactionary and outdated mechanisms of design, each one making their own personal contribution, but all equally intent upon rejecting destructive violence and exploiting the forces of irony, provocation, mysticism and poetry.

Architettura interplanetaria (Interplanetary architecture, 1970–1971), conceived after man landed on the moon, was a voyage off earthbound routes in quest of architecture unfettered by the urban nightmare, by induced needs or by planning as the only tool for regulating and solving the world's problems, a tool that for the architects of that time represented the load-bearing structure of their work.

The ship's manuals for this voyage were Marcuse's "The End of Utopia," Charles Fourier's *The Theory of the Four Movements*, R D Laing's *The Divided Self* and Roland Barthes' *Empire of Signs*. The voyagers were six strange characters, unable to drive a Fiat 500, but able to go into the cold empty spaces of the universe, each for their own reason and specific motivation, which was definitely not that of tourist observers, as had been the case with the first moon landing, on July 20, 1969, when Apollo 11 astronauts Armstrong and Aldrin went down to the moon's surface to plant the American flag and then quickly fled.

In the context of a great event, they had once again succeeded in giving a limited, partial view of the individual – a view, constructed on strictly scientific bases, of an official science that rejected utopia as a complex system of dreams and hopes, errors, memories, nostalgia, and promises of tolerance that in our belief were the basis of every architecture built for humankind. In fact, the feeling we had noticed was that of a space made alien by the new discoverers, where there was no room for the common individuals who account for planet earth's vitality.

This had been the biggest worldwide media event in the so-called civilized world, reported on the flat little TV screen, where true and false coincide and all spaces lose their emotion because everything can be cancelled out with a simple click.

However, that empty space had become a potential place of life, which Barthes likened to the emptiness of words in Zen writing: "And it is also an emptiness of language which constitutes writing; it is from this emptiness that derive the features with which Zen, in the exemption from all meaning, writes gardens, gestures, houses, flower arrangements, faces, violence."[4]

This was the place for our architecture to seek, and these were the reasons for our undertaking. The tools we set out with were few, but highly important: tools from ancient cultures modified as if by chance.

Our idea of architecture between the earth and the moon was to design without regard to fashion, monumentality, the ugly and the beautiful, the ideology of false functionality and false utility, free from the use of technology as a complex, demanding system of strict rules to be respected or else all will crumble. This idea was based on the quality the built environment can acquire when it is organized according to the use value of all things and according to the energy inherent in the mechanics of the universe. In this context, from Galileo's telescope onward, everything is architecture, not only the planets, moon, sun and stars but also the earth-moon highway, the huge transparent domes through which the stars are seen, the shifting of the planets and the footprints of visitors to the moon's surface.

The images we produced were not of regions conquered for urban development according to the ideology of the superpowers on the planet earth, but regions of the interstellar universe where architecture has existed for millions of years.

This great media event had altered the meaning of the architectural project because it had destabilized the scale of architectural relationships, the viewpoint, the perception of time and space, the depth, light, vital space, the concept of energy and especially the mode of representation. From the capsule in which Buzz Aldrin, Neil Armstrong and Michael Collins had embarked upon the conquest of space, the earth looked like a fragile little object, in continual transformation, far distant and far different from the urban settings where they lived (and from the boundless deserted regions they had traversed), which had made them believe, like all earth's inhabitants, that their planet is immense, unique and laden with infinitely recyclable energy resources. Instead, it was space that had conquered the earth, restoring the fascination of its true dimension, rather than limiting the imagination; it provided a different way of thinking about architecture, with the same visionary essence as that expressed by Fourier: "What will be precious to us will be the art of putting the heavenly bodies back to work for the creation of countermodels, in such a way that the heavenly body that gave us the lion will give us a superb and docile quadruped as a counter-model, a flexible mount, the ANTI-LION [that will enable us to travel for long periods without dismounting to rest weary limbs.]"[5]

Exceptional images began to appear, photomontages that I constructed by combining pieces from photos of diverse situations taken from everyday reality and transfigured by simple, almost invisible graphic retouches, for example *Autostrada Terra-Luna* (Earth-moon highway), *Avvicinamento dei pianeti* (Nearing the moon to the earth), *Espansione della superficie terrestre* (enlargement of the terrestrial surface), *Scene di vita sulla Luna* (Scenes of lunar life). Architecture and its representation, removed from the form- and function-bound logic of their usual confines, attained an almost unreal level of abstraction, like the theatre, cinema, philosophy and literature. Together with footage of Aldrin, the photomontages became the scenes of the film *Architettura interplanetaria,* Superstudio's first, which I directed in 16 millimetre.[6]

With our images, we intended to create countermodels that could alter the meaning of an event, by transforming the conquest of the moon from a unique, unrepeatable, purely media event into an event that belonged not to a utopian future but to the reality in which we were immersed, because theoretically the conditions of contemporary society allowed it.

Armstrong and Aldrin returned to earth carrying the moon within them, the moon on which they had walked and left their footprints forever, only to quickly abandon it. They had set out on this adventure carrying the earth within them and had eventually landed safe and sound in the waves of the Pacific Ocean, without however freeing themselves from the fascination of that abandonment. They would be overwhelmed by the reality into which they had returned and be swallowed in that ocean of things they could no longer control and guide.

Our countermodels did not manage to change the meaning of the event, but remained a sure guide for our return to earth and an important reference point against which to move, orient ourselves and continue to think of our future projects in a reality that was reorganizing itself, where the only possible environment was the existing one. The end of utopia had come in a way totally different from what Marcuse had announced.

In 1974 the work on *Cultura materiale extraurbana*[7] (Extra-urban material culture) carried out with students from the Faculty of Architecture in Florence was an attempt to study cultural processes that were not organized according to official rules transmitted through images, drawings and projects, but that occurred directly using techniques handed down in a language that did not belong to the mechanisms of official architecture. In particular the reference was to peasant culture, whose role in Italy had been fundamental in the organization, conservation and use of the land, but was now in a phase of complete transformation and, often, abandonment.

This research was important for our work, which was no longer that of designing paradoxical images as countermodels – since the society of the spectacle had now invaded the planet, including (postmodern) architecture – but of investigating the realities that had unwittingly drawn up and constructed, for their own survival, true countermodels that had become an integral part of the culture of those realities. There is an enormous heritage of knowledge to be found in this subordinate, marginal society, in which we can trace not only the roots of our science but also the possibility of a different science, an alternative way of living and planning.

The primary focus points of the research were objects – utensils, implements, methods of cultivation – various materials and assembly techniques, bindings, weaving, and the concept of use. Consuming and repurposing these elements made possible the interpretation of an identity between man and the environment and among the various stages of a lifetime. These countermodels were the object of our interest, not to copy or recycle but to understand and use to imagine a way to start over.

Zeno, a peasant from Riparbella, in the countryside of the Tuscan Maremma, is the figure who represented the exception of continuity in these cultures that were disappearing due to migration and urban acculturation.[8] He was the astronaut who guided us into this territory, on a voyage that lasted a large part of the twentieth century, during which he saw entire regions abandoned as the inhabitants suddenly left their farmhouses and all their contents: chairs, beds, tools, objects – as if they were not things that had been built with their hands and belonged to them. The peasant Zeno stayed.

His shed in the countryside of Riparbella was similar to Aldrin's space capsule. Both were self-sufficient box architectures whose external skeleton was simply a protective envelope against the space outside. Inside them, everything – floor, walls, ceiling – was used in a similar way to hang things on and from, to stretch out, eat, think and check outside through little openings, as if in either case, the sense of gravity had been suspended. The big difference was that Zeno built this "spaceship" with his own hands and could "fly" it by himself and produce the energy needed for the voyage.

He also had a house on land behind the church. It was a firmly planted building, strong and solid, with exposed stone, built by his grandfather. "This is the house where I was born on February 2, 1903, and where I want to die, where my poor papa died." There was also a large shed for the animals and a large vegetable garden in front of the workshop. It was a place where you felt safe, protected, where you could be warm in winter and cool in summer, and where the door was never locked. The workshop, however, was locked. It was where the wine and oil were stored and was only opened when Zeno returned home in the evening.

The other reference spaces of life and work for Zeno were the shed at Le Preselle and another at La Chiusa, far from the village. Zeno reached them by foot, by way of paths he had built for himself. He called them shortcuts. For him, walking was not just going somewhere, from point A to point B, but something more intimate, a ceremonial act that accompanied the beginning and end of the workday. It was not unlike the course every animal marks to identify its territory.

No division or separation exists in these cultures; everything unwinds in a continuous cycle that not only concerns life but also interior and exterior space, which is used in a total way, as if the walls were not walls, the floors were not floors, the ceilings were not ceilings. It was a space full of small objects, fragments of things, various materials placed apparently at random, that had lost their original function but that Zeno knew how to assemble, glue together and alter to create other objects or repair chairs, doors and work tools with a great understanding that came to him from his traditional culture.

It was a way of designing based on the value of reuse, executed directly without any intermediary, the project of a single maker – a way that seemed dictated by necessity alone, even if each handmade article had its own specific sign that made it recognizable, unique, precious. Their aesthetic was not based on the categories of urban consumer society, but on others, difficult to decipher and identify, and also loaded with expressive richness, pushed to the point of paradox, like in our Piper project where assemblages of objects and bicycle wheels were converted into parking spaces. It was a totally self-sufficient reality that only its maker knew how to construct, organize and consume.

Zeno's objects and utensils were paradoxes he had built for actual use and not for display, since they were a direct testimony of the creativity, use and manual ability that arise from a total self-managed relationship between the individual, society and the environment.

It was hard to find the right means for recording and the appropriate key to interpretation. Many times I entered his places with a camera or other technological device, but I almost always stopped, because I felt they were inappropriate, the wrong scale. I decided to approach Zeno using what best belonged to me, what I felt closest to and most secure with and could use nimbly; I decided to erase and redo without interference as occurs when one draws with pencil and pen, and as he himself did when he built his objects. In some cases, as I did for the shed at Le Preselle, I used collage – this time not with clippings of photographs but with bits of salvaged material. Besides drawing, I used writing, which took the place of the spoken word, the story, the reminiscence.

In the shadows of the kitchen of his house in Riparbella or in the tool shed at La Chiusa, Zeno often told stories about these objects of his, like a great planner, a great artist, a great wizard who knew how to build according to ancient principles. You set out with the desire to do, to experience, to test other paths.

1 The Piper group was composed of: R Gherardi, S Pacini, A Poli, R Russo, F Spinelli.

2 Pop art, which was born in Great Britain and had developed mostly in the United States, extended also into the world in which we were plunged after seeing the 1964 Venice Biennale, with its impressive works by Robert Rauschenberg, Jasper Johns, Jim Dine and Claes Oldenburg.

3 Herbert Marcuse, "The End of Utopia," in *Five Lectures: Psychoanalysis, Politics, and Utopia*, trans. Jeremy Shapiro and Shierry Weber (Boston: Beacon, 1970), 62–81, http://www.marcuse.org/herbert/pubs/60spubs/67endutopia/67EndUtopiaProbViol.htm (accessed February 5, 2010).

4 Roland Barthes, *Empire of Signs*, trans. Richard Howard (New York: Farrar, Straus and Giroux, 1982), 4.

5 Translated from Charles Fourier, *Traité de l'association domestique-agricole*, Volume 1 (Paris: Bossange, 1822), 529. The text in square brackets is our translation of Italo Calvino's recast of the original text in a footnote in an Italian translation of *The Theory of the Four Movements*: Charles Fourier, *Teoria dei quattro movimenti: Il nuovo mondo amoroso*, ed. Italo Calvino, trans. E. Basevi (Turin: Einaudi, 1971), 59–60n2.

6 It was followed by *Supersuperficie* (*Supersurface*), in 35 mm, for the 1972 exhibition *Italy: The New Domestic Landscape* at the Museum of Modern Art in New York.

7 See *L'esperienza "Cultura materiale extraurbana"* (Prato: Catalogo Vinci, 1977), Adolfo Natalini, Alessandro Poli and Cristiano Toraldo di Francia, "Viaggio con la matita tra gli artefatti del mondo contadino," *Modo*, no. 8 (March 1978): 49–53; Adolfo Natalini, Lorenzo Netti, Alessandro Poli, Cristiano Toraldo di Francia, *Cultura materiale extraurbana* (Florence: Alinea, 1983).

8 I met Zeno in the early '60s as I bought a house near his shed at Le Preselle, where he has his vineyard. We sold the house when Zeno died, in 1982.

Throughout the book you will find references to characters such as astronaut Buzz Aldrin or peasant Zeno Fiaschi. These are meant to be completely fictional and they refer in both cases not to Aldrin or Zeno as people but as representative of an astronaut who walked on the Moon and a peasant in the Italian countryside who sustained an ancient culture into modern times.

They are both protagonists of Alessandro Poli's fictional narration of how the Moon landing transformed his conception of our planet; other names and other people could have been substituted.

The Editors, Giovanna Borasi and Mirko Zardini

I AM INTERESTED
DIFFERENT SCALE
THE SPACE THE S
IN THEIR MINDS
DAY-TO-DAY EXPER

MICHAEL MALTZAN

N .THE RADICALLY

ETWEEN .

ENTISTS INHABIT

D THEIR

ENCE.

MICHAEL MALTZAN (born in 1959 in Levittown, New York) is the principal of Michael Maltzan Architecture in Los Angeles. He holds both a Bachelor of Fine Arts and a Bachelor of Architecture from Rhode Island School of Design, and he received a Master of Architecture degree from the Harvard University.

Michael Maltzan has created a practice that engages the increasingly complex reality of contemporary urbanization. His work with art centres, museums and housing projects for the homeless charts a new trajectory for contemporary architecture and the public realm.

From early science fiction, to the 1969 Apollo moon landing, to the present day, popular interest in space exploration has been paralleled by the production of space-related toys like model rocket ships, ray-guns and space helmets.

Discovery

Giovanna Borasi: Where were you on July 20, 1969, when Buzz Aldrin and Neil Armstrong first stepped onto the moon? Did you watch the TV broadcast of the launch? Michael Maltzan: **I did. Exactly at 3:32 pm I watched the launch. Like a lot of kids from that time, I was fascinated by the space race, and, from the beginning, with a lot of the early rocket shots, starting with Mercury, Gemini, Apollo... It was deeply a part of that culture, of being a child at that time, certainly here in the United States. Everybody talked about it. We collected astronaut cards like people collected baseball cards, and the majority of my toys were space toys of some sort; it was very much a common part of the culture. But I did watch that launch. I did watch as much of the entire flight as I could. I was, I guess, nine or ten at the time.**

Giovanna Borasi: Alessandro Poli seems to feel more of an affinity with Buzz Aldrin than with Armstrong. Aldrin had to wait a few minutes in the lunar lander before putting his feet on the moon's surface. Michael Maltzan: **I think that's true, and I think it's very interesting. Aldrin has become the more human character. He has in a lot of ways been the face of that complex relationship between making that journey and then surviving on the earth afterwards. Armstrong is such an aloof figure and so much a part of the astronaut mythology as a kind of god and super-emblem of bravado. But Aldrin, because of his own complexity, has really had a much more sustained relationship with the public. I'm sure there are plenty of personal issues going on there. It's also interesting that the majority of the images of the astronauts that we see are of Aldrin, not really of Armstrong, because Armstrong was the one with the camera.**

Giovanna Borasi: Do you like or despise the term "future"? Michael Maltzan: **I like very much the word "future" but I generally despise the image of the future.**

Giovanna Borasi: NASA's Mission to Planet Earth wasn't launched until the 1990s. The program was established because many astronauts and scientists were challenging the fact that a lot of money was invested to study other planets and systems but very little for the one we inhabit. The way I understand Superstudio's project *Architettura interplanetaria* is that it was not really driven by the idea of an expansion of geography, it's not like "conquering" or discovering another world. It was more a way of establishing a new relationship with another environment, within a new geography and a different scale. Why do you think we are so keen to go out into space? Michael Maltzan: **I don't know. Certainly, I don't know specifically why they began to refocus on the earth, but I imagine it had something to do with understanding the earth in a more objective way: the beginning of the environmental movement, the ecological movement, which was the precursor to questions about sustainability now.**

You can make an argument that the ecological movement started with the moon flights. Because it wasn't until people were able to see earth as a more finite idea that people turned their vision back to the earth. Up until that point – whether it was really about conquering or not – it was about making space a part of our context, to not just explore but to physically be there. This is an important distinction in the argument between "manned" space flight, as exploration, and science, which has never really needed people to be involved in any of these space flights. One of the interesting things about JPL – the Jet Propulsion Laboratory – is that I don't think they would ever even consider putting a person on a space flight as a part of those experiments because it adds nothing to the quantitative science. There's a different, qualitative aspect to it, though. If you look at all of the different directorates at JPL, there are people who are looking at deep space, and Mars, and other very specific areas of study. And one of those areas of study is, of course, earth science. But it always feels as if it has a peculiar position in the collective of all of these other disciplines. It's not that anybody thinks less of it, it feels like that they don't really know how to deal with it, how to characterize it. The study of earth came out of the realization that a large part of what they were doing was planetary science, which means it's kind of silly not to be studying the planet that's the most accessible to you, to understand it in relationship to those other planets. Even though what we're talking about is supposedly the quantitative component to the science and that you don't need a human being to be out there for science to take place. There is still the sense that science has the potential to operate on other more emotional level and the qualitative potential in science is possibly just as compelling reason to study space. I don't think those emotions have ever changed. That relationship between the emotional and the purely objective is probably a very difficult thing even for the scientists to understand.

between 1966 and 1968 JPL in Pasadena, California, sent unmanned spacecraft to both the moon and Venus. Groups of Apollo astronauts visited JPL to look at photographs from the Surveyor missions, which showed that they would be able to safely land and walk on the lunar surface.

Giovanna Borasi: Current scholars, such as Volker M Welter, Vittoria Di Palma and Robert Poole are now going back to the 1960s and the 1970s lunar missions to develop the idea that leaving the earth's atmosphere was in fact the only way to discover and understand better who we are, and what is earth. What these missions brought back, in fact, was the first image of the complete earth. I think it's interesting that in a space mission, we go out there to

One of the first missions to use the Space Flight Operations Facility at JPL was Ranger 7, which went to the moon in July 1964. In 1985 the SFOF was designated a National Historic Landmark.

discover something else, but what we actually discover is where we come from.

Michael Maltzan: **That's one of the things that's been interesting to me about the Poli research: this very complex relationship between what it would mean to be in space and what it means to somehow return and be on the earth.** No matter how far the experiments are out in space, no matter how objective they are about looking, very, very deeply in many cases, into space (and in some of those missions, the probe is still on its way; it will continue to expand the range where data comes back to us from), it's always contextualized by the fact that the scientists themselves are here. They are not physically leaving their day-to-day experience, and so after you've spent an entire day with that information, and have put yourself somehow in that context of being that removed from the earth, you still leave, go out the door, get in your car, drive home, make dinner. The radicalness of that kind of juxtaposition is interesting. I've never heard them talk about it. But it's there in the sense that, for us, there were constantly two currents in designing the JPL building. One was the work that they were doing, and the other was the culture of the place. And the culture of the place is incredibly recognizable as any kind of office culture, with all of the politics and the social dynamics that exist in those cultures. But there's also this strange sense that at any moment that conversation can instantaneously elevate to a very different plane of understanding, a very different plane of thinking. And the elasticity with which they seem to move between those two worlds, I think, is very interesting.

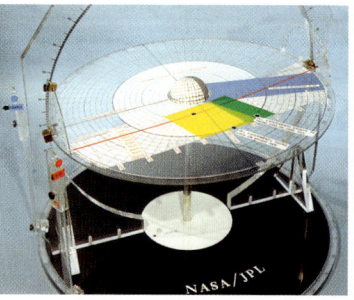

A 1963 issue of *Lab-Oratory* asked: "Ever try to shoot a slow flying duck while standing rigidly on a fast rotating platform, and with a gun that uses bullets which curve 90° while in flight?" This 1967 trajectory design model allowed mission planners to illustrate the orientation of Mars, the expected path of the Mariner 6 and 7 spacecraft, and the window of opportunity for operating instruments and TV cameras.

A prototype device was developed by JPL in 1969 to split lunar rocks into pieces and then pressure-weld lids onto sample containers. It had to fit in a 30 cm cube, be easily moved and disassembled, and made of aluminum and stainless steel, which would not contaminate the lunar samples.

Giovanna Borasi: In this sense, the JPL approach, and this idea of sending instruments rather than humans on voyages of discovery, is very interesting. Intuitively it seems to me that it differentiates the scope of the research from the approach in which you would send a person to physically go somewhere. Two different ideas arise: bringing back data versus bringing back the *experience* of that place. These days they are bombarding the moon's surface to find out if there is water. At the same time scientific exploration is only one way that people are interacting with outer space, as there are also all kinds of other different projects happening, with different motivations. Space tourism, for example: it seems to me more like

Mariner 4 was launched in 1964 and provided the first close-range images of Mars. A machine converted digital image data into numbers printed on strips of paper. Too anxious to wait for the official processed image, employees attached these strips side by side to a display panel and hand-coloured the numbers like a paint-by-numbers picture.

In 1975, an 11.5 × 2.5 m panorama photographed on Mars by *Mariner 9* was coloured by Don Davis, an artist with the US Geological Survey, as the background for the new Mars Yard, the first of many created at JPL for testing Martian vehicles.

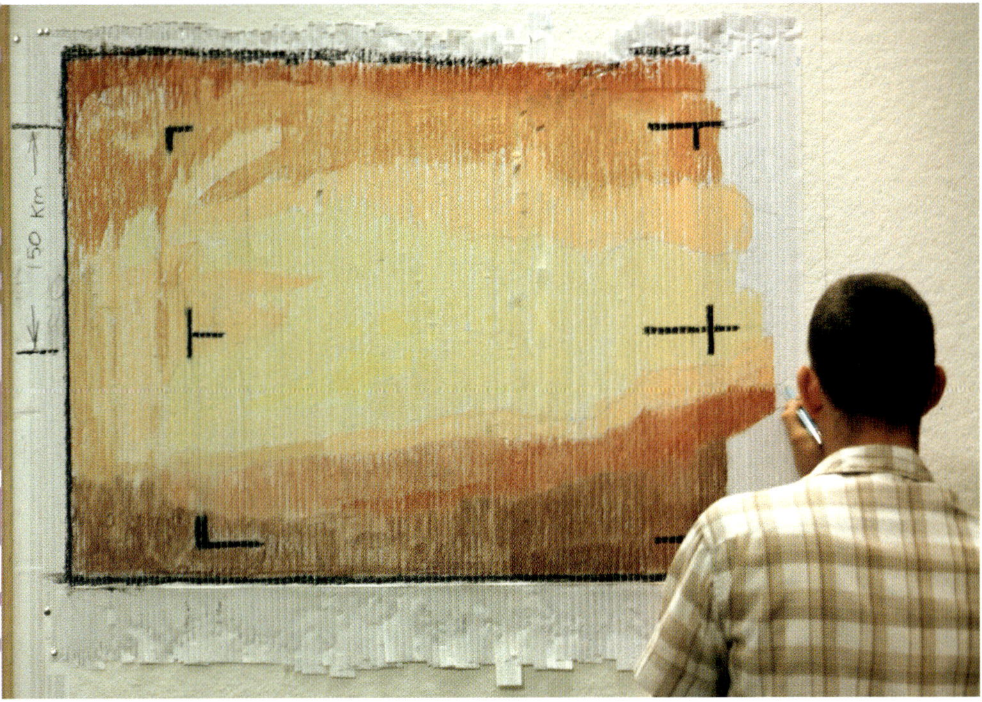
150 Km

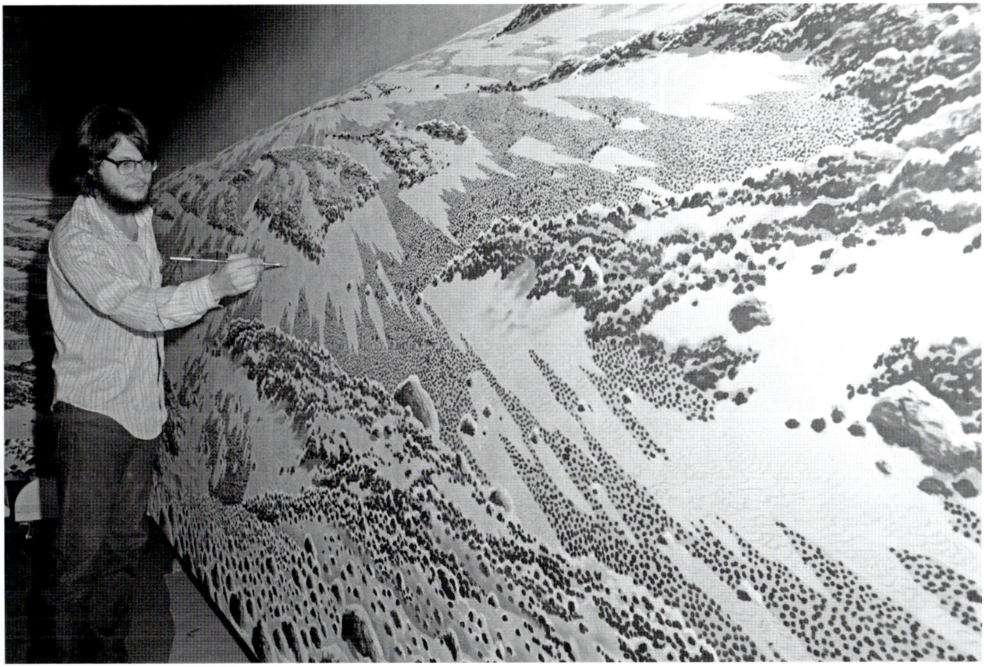

In 1965 JPL was working on methods to enable a spacecraft to safely reach the surface of a planet. One of these was an "impact limiter" that would cushion the landing. The prototype impact limiter included a series of Mylar "convexities," Mylar/nylon webbing, and cords made of music wire.

The JPL-managed Seasat mission, launched on June 26, 1978 using an Atlas-Agena rocket, pioneered satellite oceanography with this synthetic aperture radar antenna. For over three months, until its power system failed, Seasat observed sea-surface winds and temperatures, wave heights, internal waves, atmospheric water, sea ice features and ocean topography.

the next generation of Luna Park, in a way. It is really about going out from our atmosphere for some time and then coming back, and I imagine it's about experiencing that feeling of abandoning the earth. The discovery is ultimately a personal experience. Michael Maltzan: **Just think, we're at war with the moon!** **I think there is a difference between space tourism and the way in which NASA's space flights relate to culture as a whole. When it was just that group of astronauts, it still had a meaning, because it was a corps of explorers that were going to these places, supposedly in the name of humankind as a whole. At least, that was the romantic notion of it. And because of that, it was possible for all of us to very much be a part of that as a culture. While we weren't physically going there, they were our satellites in a way, they were our research instruments. They were the eyes that somehow brought back the images to us, and produced something in a collective imagination of what it meant to actually travel in space. What JPL is doing with their space flights (the satellites, the instruments, the experiments) is very much that. They are a kind of peculiar mirror: even though they're looking very far out, what gets returned is something that is collectively held. The incredible bank of images that NASA is now sitting on from these experiments, of planets, galaxies, of the exploration they're doing, really does continue to be in a real and expansive relationship to culture as a whole.** But **it's impossible not to think of space tourism as the jurisdiction of fabulously wealthy individuals, so that collective imagination might start to disappear as it gets replaced by the experiences of a few. It doesn't feel like any of that space tourism so far has resulted in an expansion of how we all are a part of that experience.**

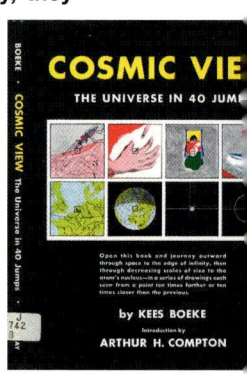

Published in 1957, Kees Boeke's *Cosmic View* illustrates the relative scale of things in the universe with drawings that zoom at an exponential rate from the surface of the earth to the outermost reaches of the universe and then back down to the inner workings of the atom.

Giovanna Borasi: If public space travel becomes a commercial reality, what role can architecture play in the way we imagine new opportunities for living? Will this new reality be an inspiring thing for architecture? Michael Maltzan: **It's possible. But in the beginning the technical challenges, I imagine, will be the most prominent thing in those projects. And much of the role or responsibility of architecture, in terms of its connection culturally, seems in that context likely to be "extra," not particularly necessary for what that kind of space travel might actually be. I would be less interested in architecture's relationship to producing some kind of way of living in space, and much more interested in what that would mean to architecture here on earth. The idea of an expanded habitation, the idea of a changing balance of what is finite on the earth, a changing balance in terms of limits, a changing balance in terms of the culture of even communication and connection, the difference in temporality – all this would be very much a part of the conversation of both living and going into space, and would likely have an effect on the way that we think about architecture here.** I think that that has some equivalency to the work that the scientists at JPL are doing. Again, information and images that are produced very far away have potentially a big effect on the way that we think here

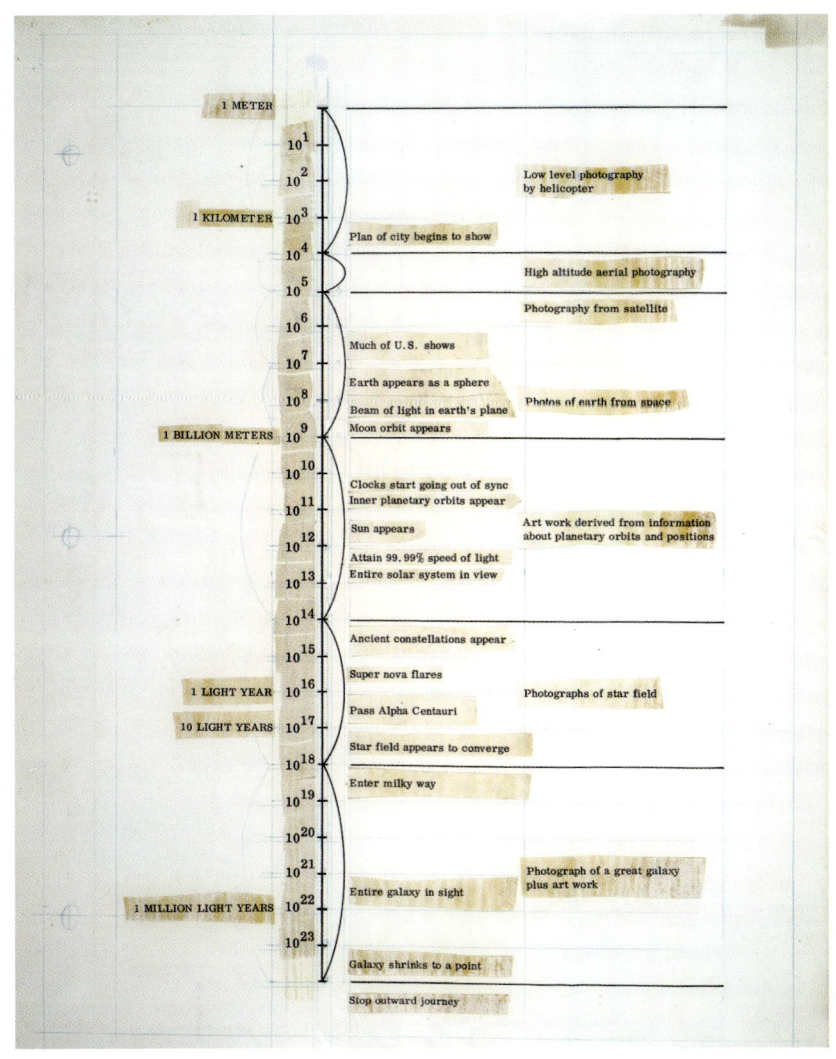

The 1977 Charles and Ray Eames film *Powers of Ten* (inspired by *Cosmic View*) communicated jumps in scale through low level earth images taken by helicopter, high altitude views captured by satellite, the whole earth photographed from outer space, and planetary orbits through artwork based on scientific data.

conceptually. Our potential for abstraction and imagination might radically be transformed.

Giovanna Borasi: When the new building for JPL was commissioned, what was the first image that came to your mind? Michael Maltzan: **It wasn't so much a first image, it was really a first dilemma, the dilemma of scale. This sense of how radically different the space that the scientists inhabit in their minds is from the day-to-day experience that they have just working in their offices. I couldn't understand how they navigated that difference. When I was working on a concert hall a number of years ago, I kept looking at the ac- ousticians and wondering. I had this fantasy that their ears were larger somehow than anybody else's, and I kept staring at their ears to see if they were a little bit larger. And in the same way with the space scientists and** the rocket scientists, I keep won- dering if there's some part of the way that either their bodies or their minds work that allows them to make those kinds of scale shifts that just seem abstract for most of us. You can quantify it, you can understand the numbers, but your ability to actu- ally feel that even emotionally I think is impoverished. And yet, they seem to be able to be much more con- versant with that. I've never gotten a clear sense from them whether that is because they've just gotten

In his 1969 series *Yucatan Mirror Displacements*, artist Robert Smithson put mirrors on the ground and in the trees throughout the Yucatán and the photographed the resulting inversions of sky, land and earth.

used to it and they don't really question it much anymore, or whether they've adapted in a very particular way to allow them some kind of emotional connection to that question of scale.

Giovanna Borasi: I feel that you are describing a particular approach to the question you are dealing with. In this sense, I think that with the JPL project you chal- lenged yourself with the idea of designing a building that will incorporate this special attitude of jumping from one scale to another or the capacity to see at the same time our planet and "out there." I was in Pasadena and drove to the JPL campus and pulling up in front of the buildings I had the feeling it was nothing special. It gives the illusion that there isn't anything adventurous going on there. It seems like any group of office buildings…

Michael Maltzan: **Even from close up, it's the most perfunctory of campuses you can imagine.**

Giovanna Borasi: I'm curious. It seems to me that you felt challenged by this fact, in wanting to propose a building that has some additional values, not simply delivering them some offices.

Michael Maltzan: **I think that's true. I hope the building operates on a number of different levels, and I think reimagining a way that the office culture could potentially work, and searching**

Johan van der Keuken's 1975 photograph *Mountains Outside/Mountains Inside* appeared in Herman Hertzberger's 2000 book *Space and the Architect: Lessons in Architecture 2*. For Hertzberger, this image depicts a mirroring of experiences inside and outside, showing how an external landscape might visually enter an interior space and then be projected into the mental space of metaphor and memory.

for a way to create a much greater level of interactivity to break down the hierarchies which have grown up culturally at JPL over time, is part of the responsibility or the role of the building. But it wasn't the important part of the project. That was important to the campus, and that was important to that particular culture. But it didn't seem to me to be necessarily the resonant thing for the scientists themselves. It is true that that campus and the majority of the buildings, especially externally, are deeply mundane, and maybe the landscape there and the outside spaces are the better part of the campus. But, regardless, I was much more interested in how the experience of being in the building had some kind of equivalency to the very abstract characteristics of what they were involved in. I wasn't trying to represent the context of space, or the specific iconography of space flight, or to allude to the form of the things that they make there – satellites and instruments. I was trying to create an equivalent at some level to the conceptual, abstract space that they live in because I felt that if I could do that, the building and the space would have a greater level of connection to that ability to mediate somehow or to live in the space between those radical scale differentials. The building is not trying to *mediate* those experiences, the building is trying to be the *space* of those experiences.

Giovanna Borasi: You were telling me yesterday that in a meeting with JPL scientists and administration they were looking at your different models and proposals, and they started to discover in some of them forms that reminded them of aspects of the things they do. Your reaction was to change your approach to the project, because you didn't want to create a replica of what they are doing. I get the impression you decided to design a building that they could relate to and that could facilitate what they are doing on other levels.

In her 2004 painting *Stadia II*, artist Julie Mehretu take urban architecture as a point of departure to convey a compressed sense of time, space and place.

Michael Maltzan: Given the constantly evolving and expansive quality of what they do, you almost want to make a building that is constantly expanding as well. But it doesn't necessarily expand physically. In fact the form is finite – and needs to be as a building. So the question is, how can that constant reinvention or expansion happen within the surface of a finite form? That's very much the quality that we're trying to create on the inside of that building. The plan in certain spaces is absolutely fixed; it's not about flexibility. It's about a very abstract way in which the space has a level of conceptual elasticity, to keep up with or to change as the culture of that place continues to emerge, as the knowledge, the intellect, the intelligence of that place continues to emerge.

Giovanna Borasi: I'm reading *Biomimicry: Innovation Inspired by Nature*, by Janine M Benyus, who says, "We have a habit of making theories about organisms basing them on the machine of the hour." She observes that people always relate contemporary reality to what they think is the highest achievement in knowledge and technology at that time. And so people might say that the human body is like a clock, or our brain is like a computer. It seems to me that there are a lot of JPL buildings that reveal in their architecture the scientific ideas of the period when they were built. Michael Maltzan: **If you look at a lot of the facilities that house NASA and the space program, those buildings try to, in very traditional architectural terms, represent some kind of form that's equivalent to our sense of the space program. Maybe the best of those are really the ones that are filled with a kind of bravado. They're monumental and iconic at the scale of some of the largest parts of the program.**

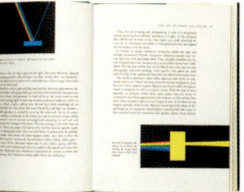

In her 1971 book *The Splendor of Iridescence*, Hilda Simon illustrates colours in the animal world through drawings of shifting iridescent scales on the wings of blue morpho butterflies and diagrams of the visible and invisible spectrum.

But that's not what this project is about – at all.

Giovanna Borasi: You often describe the building using analogies with scientific ideas, like "the surface of last scattering" or "the spectrum of chroma"… So I'm wondering if you borrowed these concepts from other disciplines to describe the project or if you used these concepts to organize and structure the project in a different way? Michael Maltzan: **One of my biggest fears was that we would engage in a kind of pseudo science, that we would latch onto certain ideas in the most pedestrian of ways and use them primarily in analogous or metaphorical or representational terms. And while we've used some of those ideas that have a relationship to the space and planetary science that's being done at JPL, the real goal was to use them as starting points for ways of describing much more abstract characteristics within the building, socially, certainly spatially, and, at that level, to some extent in terms of the form. But we were trying to produce an experience and an atmosphere and a set of qualities more than, say, this facade represents this particular scientific idea. Fundamentally, this project is about the space of the mind, not the space of the body. If there is an overarching idea, it's that. And all of those things like chroma or optics, they certainly have a physical effect, but that's not what they're really there to produce. They're really there to produce a much stronger characteristic space of the mind, in terms of things like creativity or imagination, certainly, but also in terms of the openness and the sense of the possible that comes from the abstract, from formlessness. It's like the Julie Mehretu painting. What I see in it is a kind of form that is created by particularization of elements; there is clearly a sense of movement and gesture, and potentially the form of a spiralling in that image. What's really important to me in that is the allusion to the beauty in equivalence and the non-hierarchical, and that to me is a space of the possible, of ideas. And it's a space that finds a way to fuse the radical *intellectual* break which we need to make to imagine scale at the level that these scientists are dealing with, with a kind of beauty that relates to the *emotive*, qualitative**

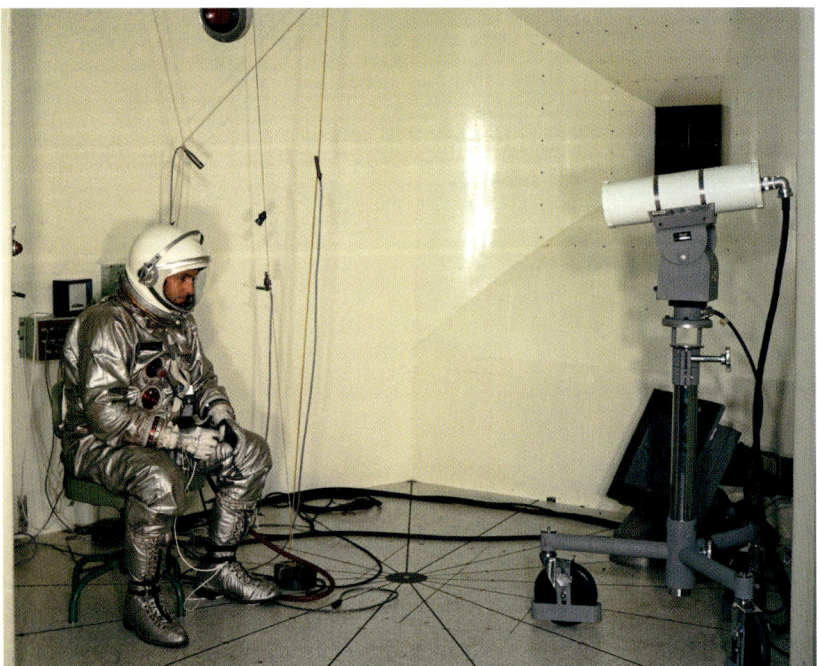

In 1964 the acoustic reverberation chamber of the JPL Environmental and Dynamic Testing Laboratory in building 144 was the site for hearing tests, which were done to verify that astronauts of the Apollo program could hear a warning tone above the noise generated by the huge Saturn C-5 launch vehicle.

aspects of what they do. They have to be passionate about their work to be involved in something for eighteen or twenty years! A single idea! Their motivation is not just intellectual curiosity, although I think they have that in extraordinary amounts. But it's somehow the beauty in trying to find some level of meaning in these very abstract, almost impossible spaces of inquiry. How do you continue to find meaning in that when everything that you're confronted with is, in a lot of ways, the intangible? There are very tangible things that you have to do in the science to set up the experiment, design the instrument, launch the satellite, and continue to communicate with it and get the communication or images back. But that's the mundane part. It might keep you motivated on some level, but what actually sustains you? It's this other emotive quality.   What I'm trying to get at is that I'm not interested in this because it's fantastic that we got a project about space. Here, the problem is that that experience and what it's related to is something that is so abstract and so out of scale that a lot of the normal devices of architecture don't work in relationship to that. A lot of the classical devices of architecture seem minor in comparison to the scale of the journey that we're actually talking about. Even though, ironically, it's completely land-based – physically.

In June 1955 a JPL employee wore an "arctic suit" for tests at −40° in the Environmental Testing Laboratory temperature chamber, a space capable of reaching temperatures of −75 to 150°C and used to simulate the temperature extremes of outer space.

Giovanna Borasi: Does this mean that this project gave you the opportunity to explore new devices, new ways of defining architectural elements? For example, the structure, the facade, etc., to have new qualities? Michael Maltzan: **Well, think of the skin: we've talked about it in terms of qualities of scale. Its non-hierarchical qualities and its peculiar scale qualities are purposeful, because we didn't want to be able to read the conventions of building construction in that facade. For instance, if it was a glass facade you'd see mullion spacing; the dimensions of that facade would have produced something that you had an instantaneous relationship to. So, in that sense, there are qualities that eventually become the architecture.**

Giovanna Borasi: If we return to the concept for the building as the space of the mind, you said that one of the qualities it has is that it's not hierarchical. Michael Maltzan: **Take, for instance, the surface of last scattering and the surface of the building. I would say that this concern with the non-hierarchical and form-lessness as it relates to space has as much to do with not composing the building from the standpoint of very specific limits. The boundary of the building wasn't produced as a form specifically. It is, again like the Mehretu painting. For me, I could see this building as smaller or significantly larger. Where the building starts and stops is nearly, literally, the moment that the building stopped, and it is an extremely thin line that describes a surface, a boundary. But I wasn't really trying to produce a form from the beginning. And in that sense it's a kind of snapshot at that particular moment. That's one of the reasons why we've tried to produce as much thinness in that surface as possible. And why some of the more defining spaces are not completely internal to the building, but are in a somewhat uncomfortable position on that surface. We went around and around on those spaces trying to imagine them not so much as complete specific rooms, as having an in-between quality between the outside and the inside. I was thinking – you know colour space gamut diagrams? In graphics and photography, you have these colour spaces, and they have different gamut sizes. You have the full spectrum, and it's actually quantified by its size, and then there's always a kind of gamut… the funny shape, that is very thin, in a way. It's a very abstract thing, but that's literally the form of the colour space. That to me feels equivalent to what we did with the building and its relationship to the different parts. All of these things were a way of producing a vocabulary that wasn't specifically architectural, and that became almost a kind of shorthand that referred to one part of the design strategy.**

Giovanna Borasi: So you had to define a new vocabulary to deal with some issues because of the specificity of the nature of the project. Like Greg Lynn's and Alessandro Poli's research, it deals with the idea of imagining a new frame of reference and building a new vocabulary for it. So, maybe this is the right way to describe your project, as "a space of the mind." The exhibition title is *Other Space Odysseys*, and within the ambiguity of the word "space," we will perhaps discover the very different journeys that you will all bring us on.

Michael Maltzan: Well, that sense of journey absolutely is a part of what we're talking about here. It's just that the journey of those scientists in that building is not really a physical journey. It's not the kind of journey that has a tangibility that human space travel has. But I think it's no less prominent, and in a lot of ways is at times that much more captivating, that much more intense, because it is very much this equivalent journey that you have to produce in your mind that collapses all of these dimensional aspects that generally define "journey": scale, time, all of the reference points. All of those things are completely fluid and elastic in the mind in a way that are not as elastic or abstract if you're taking the actual journey.

Giovanna Borasi: The last question: what is your aim with this exhibition that we are doing together? Michael Maltzan: I think we started by trying to assemble what the particular themes are that relate to the building. When we started to disassemble these themes, the focus began to be not so much the story of the physical attributes of the building but these very abstract qualities, and this is the part that I think is a little bit more difficult from the standpoint of the exhibition. I would be very happy for someone to *not* understand, necessarily, the building, which is a risk. I started to imagine that if you could create this project that was based not so much on a totalizing structure, on neither the beginning nor the end... but somehow the in-between, that would have an equivalency to what we were trying to produce in the building in terms of the experience. I think it is better if the show provokes you to ask questions as opposed to providing any. Because the building as a whole provides an answer. It's the ending to the story. So, it's probably good to not have the ending.

JET PROPULSION LABORATORY, PASADENA:
A NEW BUILDING

Michael Maltzan Architecture

The Jet Propulsion Laboratory (JPL) was established at the California Institute of Technology in the 1930s. Since launching America's first satellite, Explorer 1, in 1958, the JPL has pushed the outer edge of space exploration through research and innovation. JPL's robotic craft landed on the moon in preparation for the Apollo missions and continue to survey the solar system and beyond.

JPL is a NASA laboratory whose research includes studying the nature of the Martian surface; the causes and mitigation of ozone depletion and global warming in earth's atmosphere; and the search for life in and the nature and evolution of the universe. Its missions include the Mars rovers *Spirit* and *Opportunity*; the Mars Reconnaissance Orbiter, which since its launch in 1996 has returned more than 240,000 images of the Red Planet; Cassini, currently exploring Saturn and its moons; and *Voyagers 1* and *2*, which are now beyond the edge of the solar system. JPL also manages NASA's Deep Space Network, an international network of antennas on several continents that serves as the communications gateway between distant spacecraft and the earth-based teams that guide them.

The design for JPL's new building creates a dynamic, interconnected organization, which reflects and fosters the inventive spirit of the institution it represents. The building will house the executive leadership of the Jet Propulsion Laboratory, including JPL's Executive Council, directors and deputy directors, and key scientific and administrative staff. Addressing the interrelated programs of scientific research, interplanetary communication and technical innovation undertaken by JPL, the design proposes a new architectural model for scientific research.

The project departs from hierarchical structures that typify office buildings, instead constructing a series of flexible relationships – in both organization and morphology – that catalyze interactions between the building and its context, its form and use. In this way, the design for the building seeks to transform the landscape of the workplace, encouraging a more abstract, conceptual space for experience, a space for new possibilities and connections, a space of the mind.

The project poses an alternative to typical office planning, with the traditional centralized core distributed across its floorplate. A series of flexible, double-height spaces spiral upward from the building's base, integrating areas for singular, focused study and zones for more informal connection and collaboration. This array of spaces cuts across and through the building's polygonal form, linking the building's interior landscape and the campus context into a new, otherworldly topography of movement and perspective. Challenging existing paradigms for the workplace by creating a new network for interaction and engagement, the design informs and supports the JPL's ongoing mission of research and innovation.

An aerial view taken January 1, 1961 showing the JPL
and the surrounding San Gabriel Mountains in Pasadena.

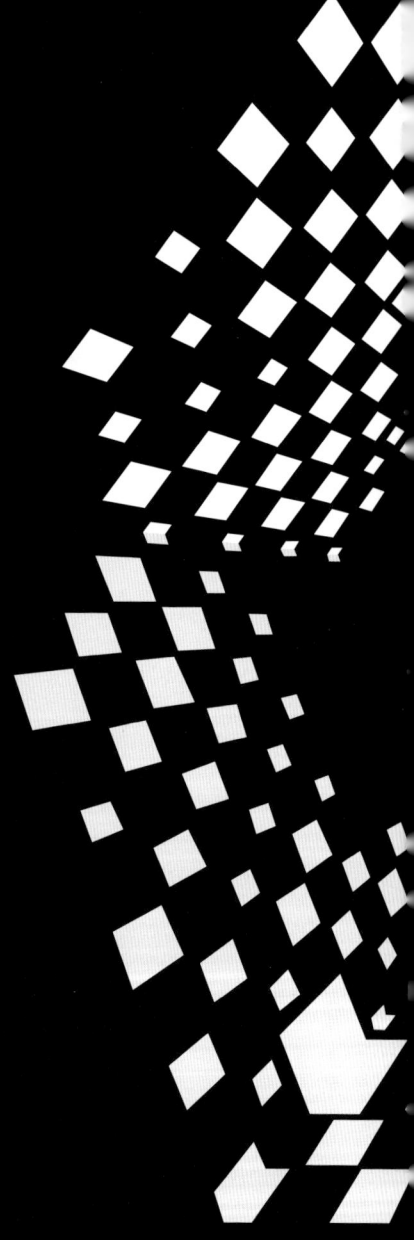

New building for the Jet Propulsion Laboratory, Pasadena, California, Michael Maltzan, 2006–.

Right: Perspective from inside of the building envelope.
Rendering
Page 142: The exterior envelope of the building.
Model: Laser-cut museum board
Page 143: Possible variations for the texture of the envelope, conceived as a membrane with openings that vary continuously in size.
Digital drawing
Page 144: Conceptual organization of the building plan with a compression in the space organization on the diagonal axis.
Model: Foam board and acrylic
Page 145: Relation of the inside plan organization with the exterior envelope.
Model: Foam board and museum board
Pages 146–147: Axonometric studies of variations for the form of the building.
Rendering

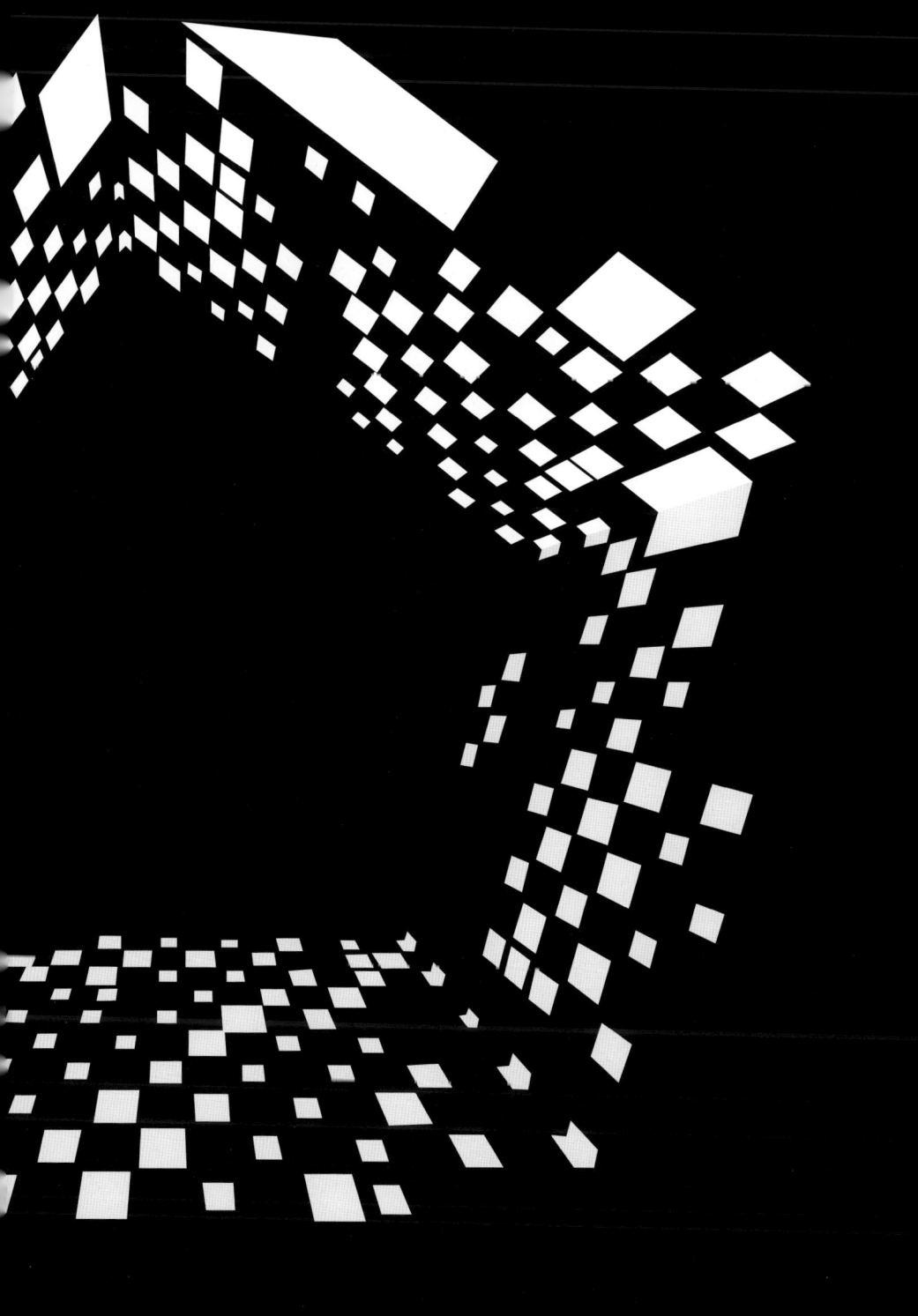

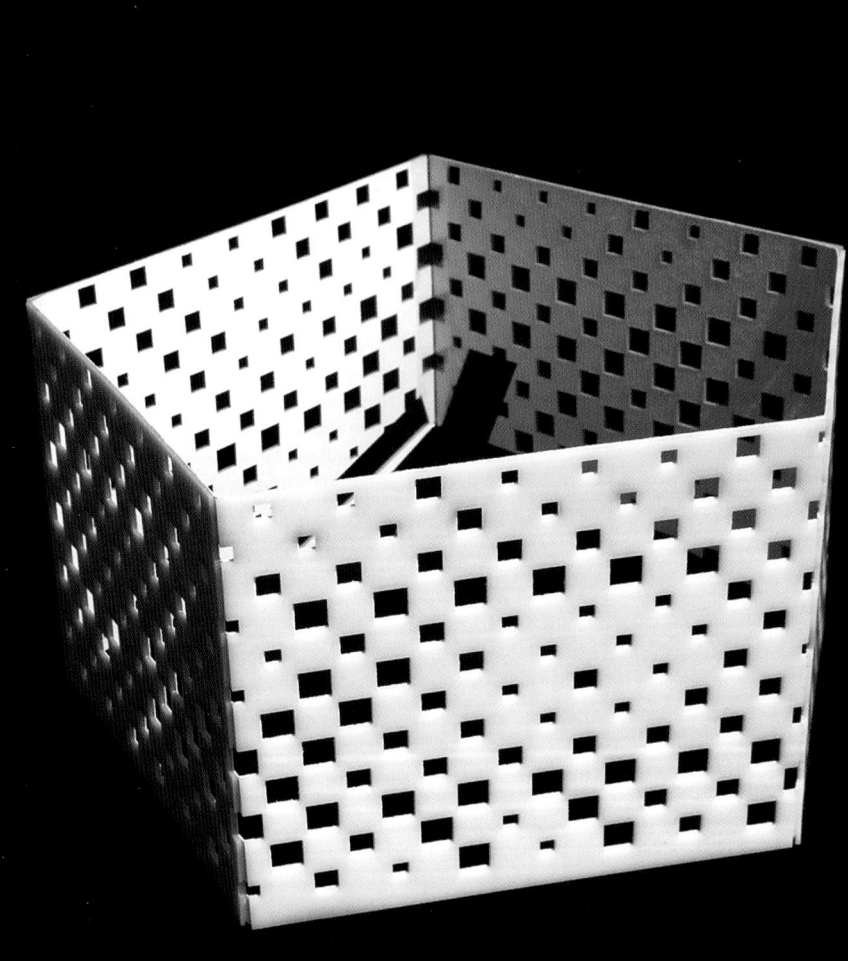

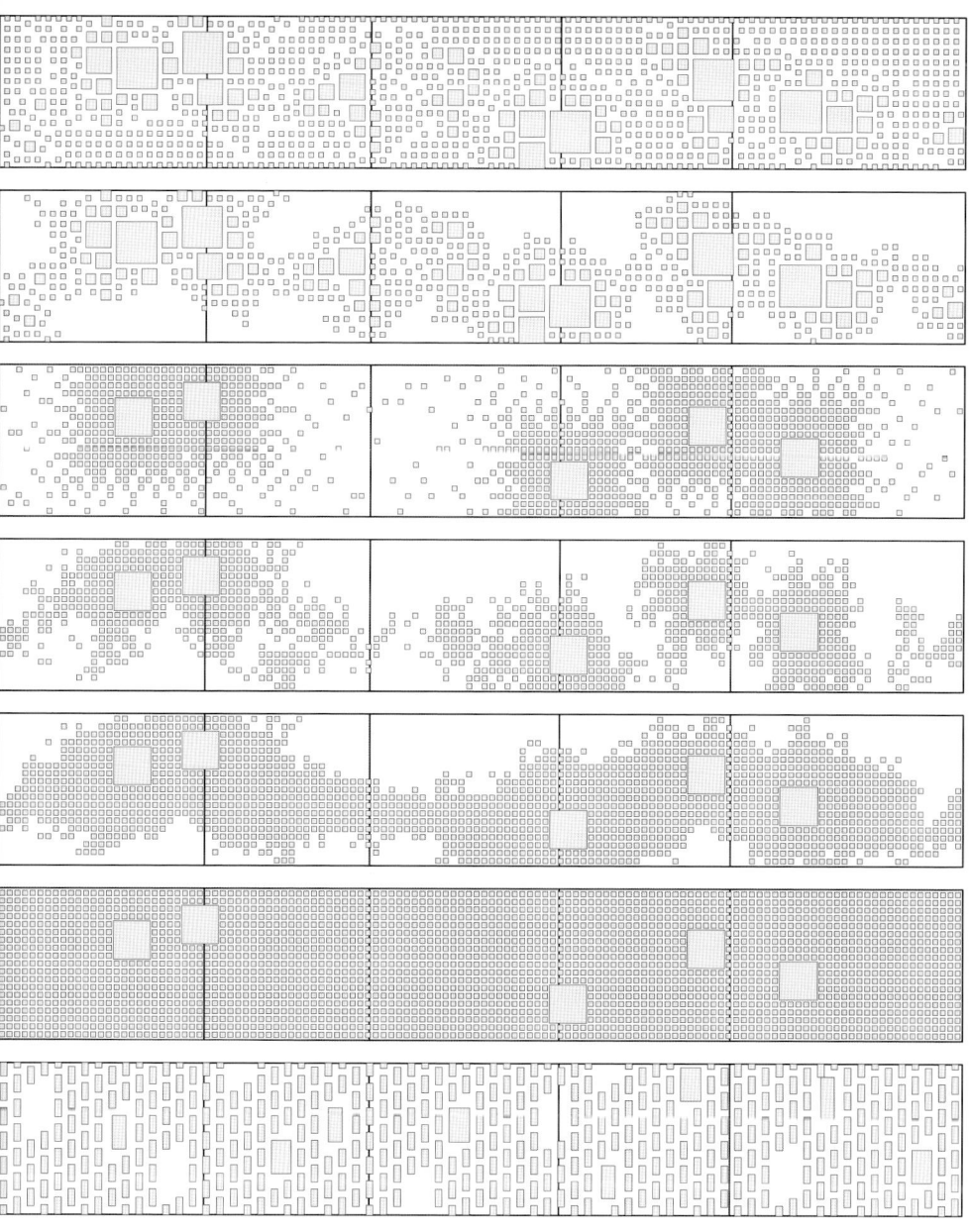

143

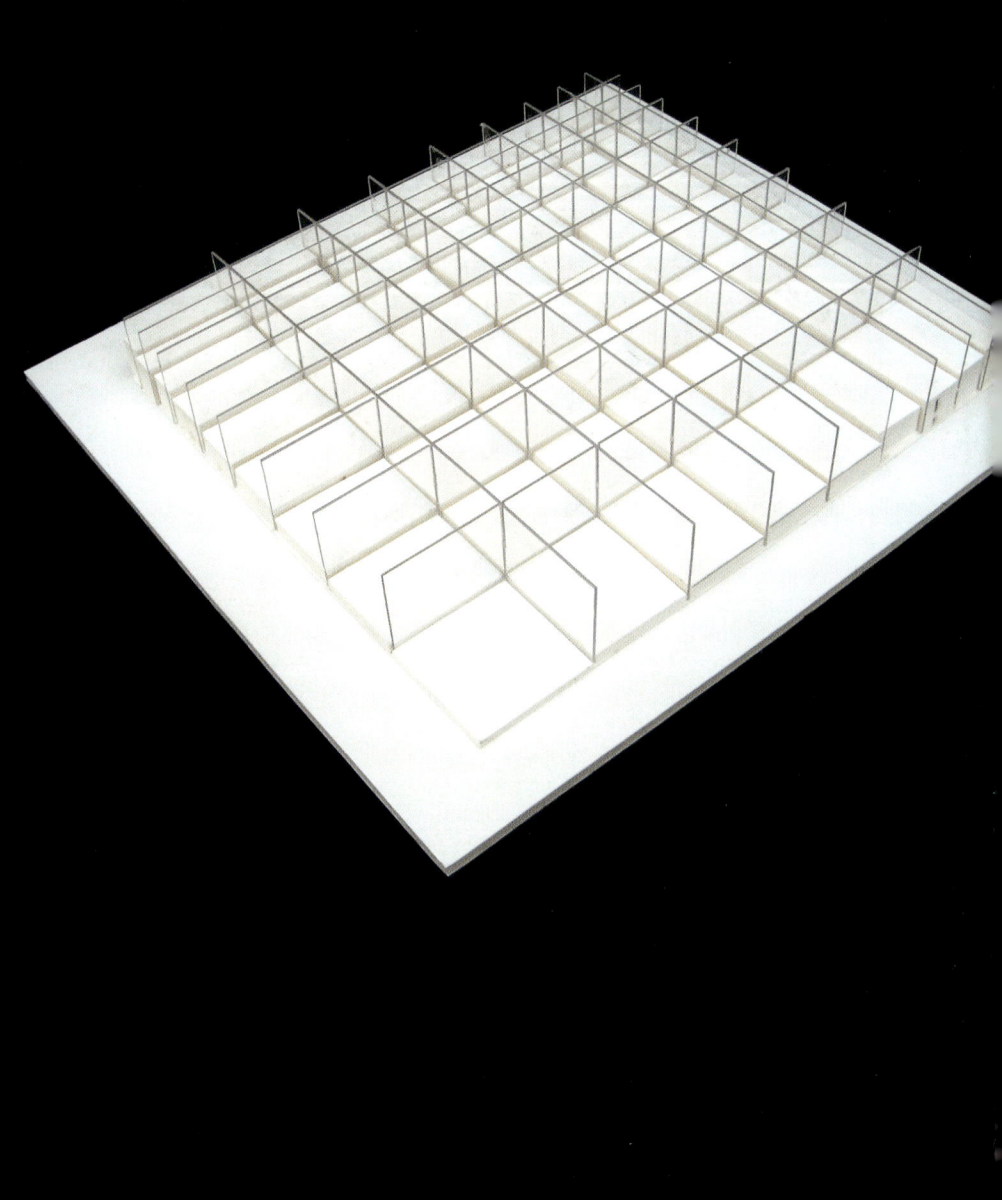

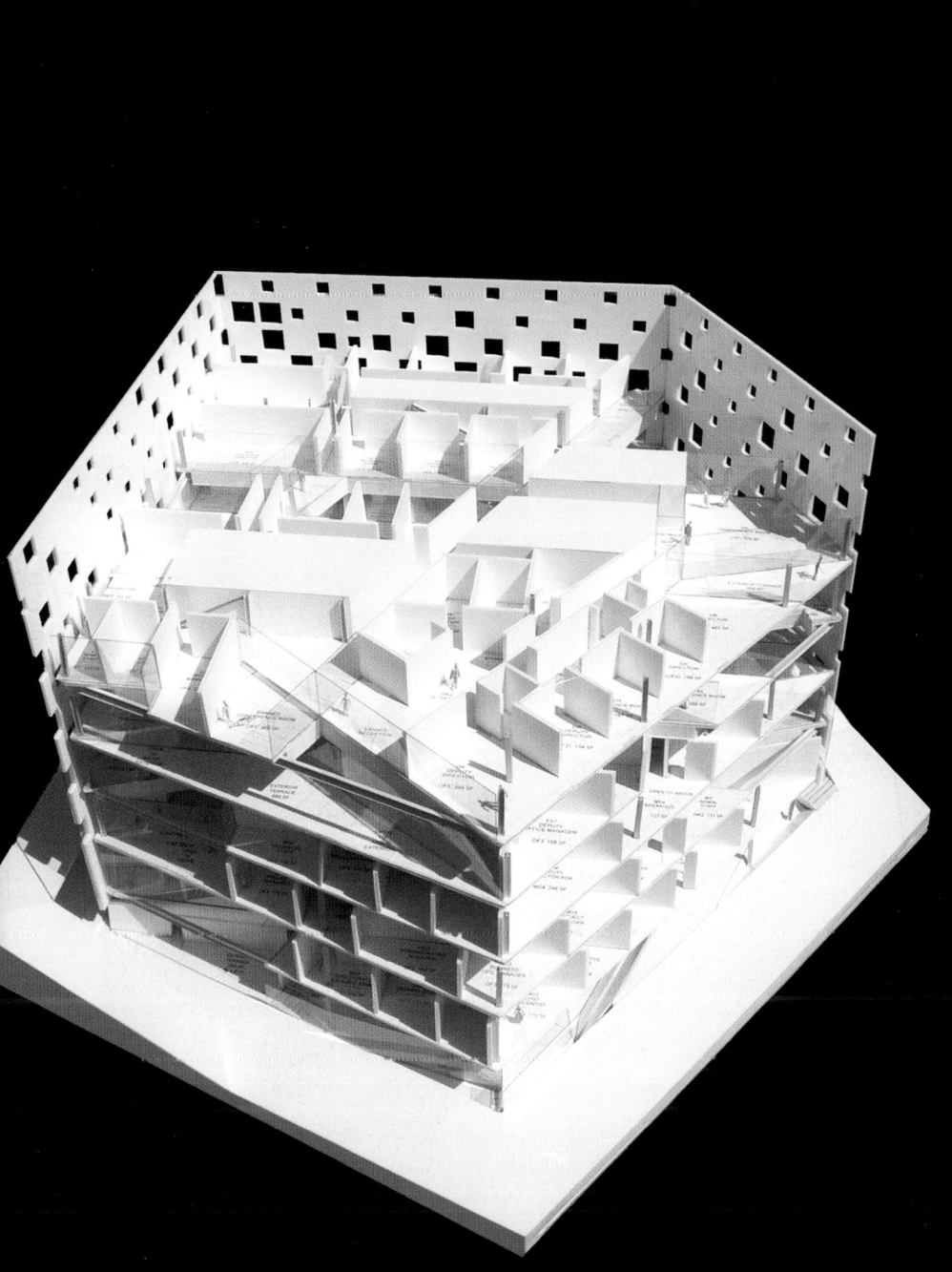

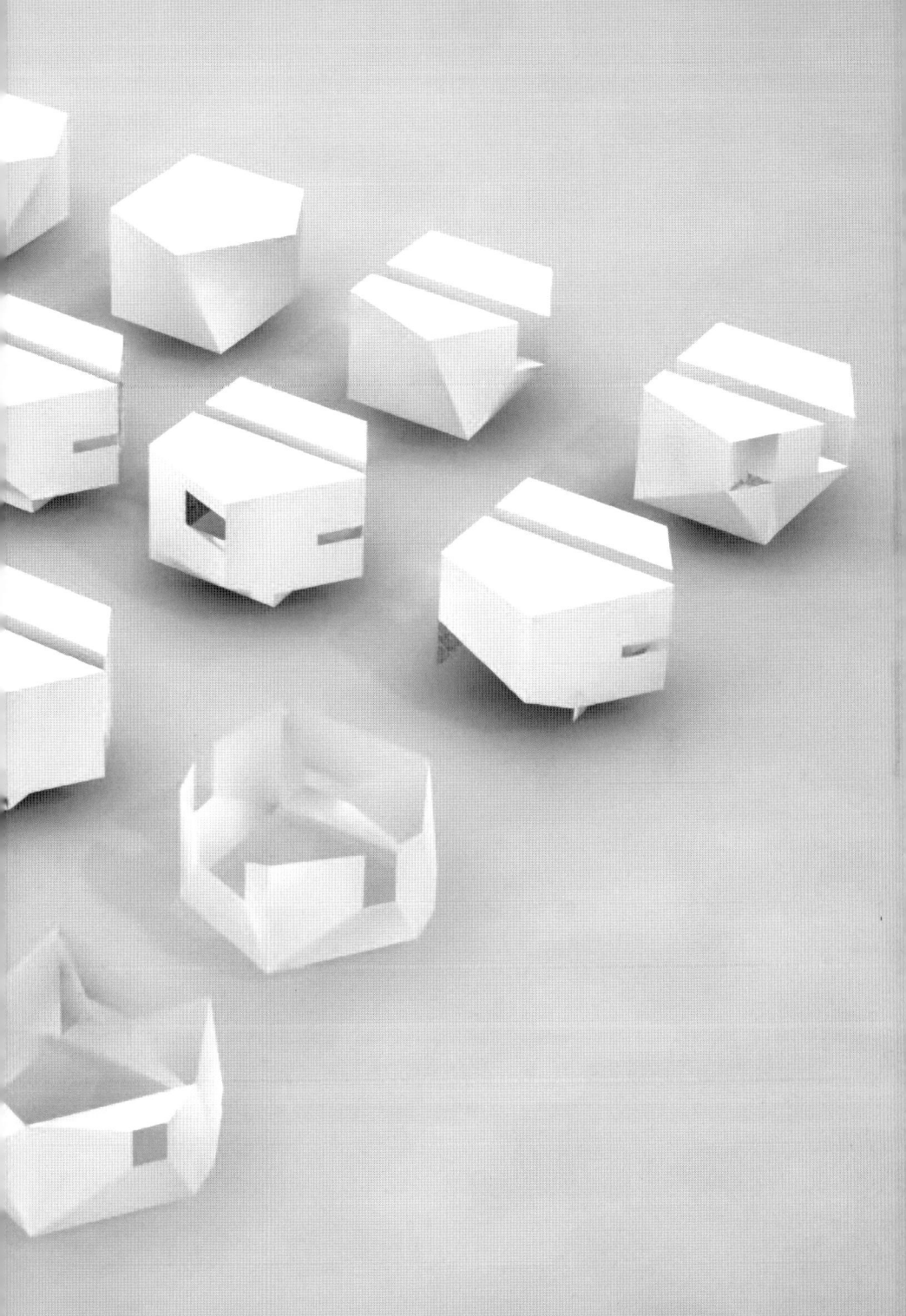

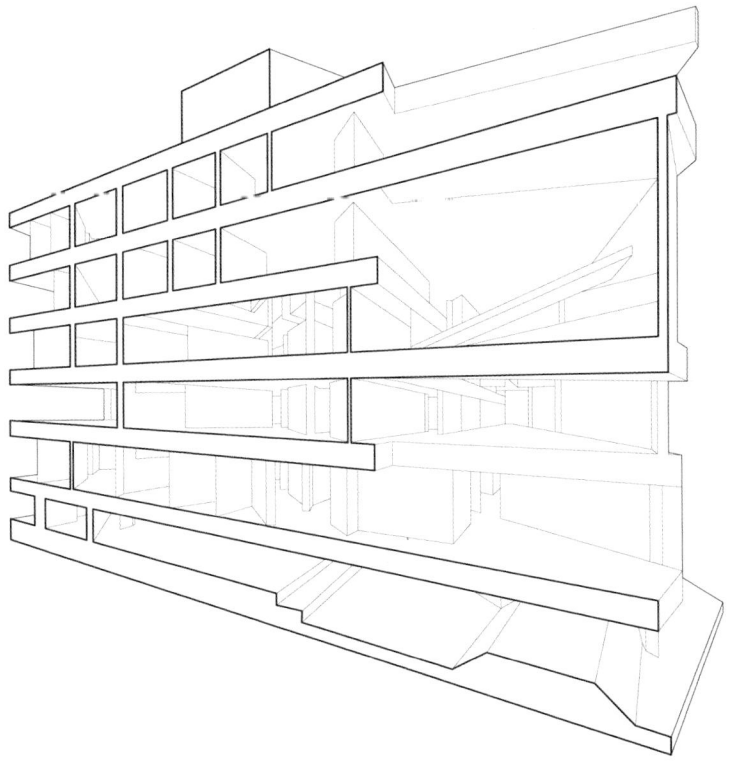

New building for the Jet Propulsion Laboratory,
Pasadena, California, Michael Maltzan, 2006–.
Axonometric perspectives: small and large double-
height spaces are distributed in the plans to
challenge the traditional hierarchical organization
between the different functions of working spaces.
Digital drawings

Pages 150–151: Detail of the facade with openings
varying in size.
Model: Laser-cut acrylic, foam board, gator
board, wood

BEING ZENO FIASCHI TODAY

Mirko Zardini

Although books about exploration celebrate the moon landing as a heroic culmination in the human saga of conquest, space exploration does not have an important place in canonical history books – even those that focus on the second half of the twentieth century. When the subject does come up, it is at most as part of a chapter dealing with the Cold War rivalry between the USA and the USSR. Nor does the conquest of space generally figure in historical atlases that interpret the narrative of events through maps. And space exploration has had limited economic impact, as historian Gabriel Tortella has noted.[1] Besides a few years of collective euphoria, it seems to have had no particular effect on our daily lives or the events that have marked our planet. This chapter in history may be considered closed.

Today's new interest in outer space is due to factors different from those of the 1960s: This time the political and military rivalry is between the USA and China, but, more importantly, exploration today is driven by economic motives and the establishment of a new form of tourism – space tourism. Perhaps the coming decades will see man no longer just orbiting the earth, but again landing on the moon or embarking upon further explorations.

The return of the last astronaut from the moon on December 19, 1972, and the energy crisis in the winter of 1973–1974 marked the end of what economists like to call the "thirty glorious years"[2] that followed World War II. At least in the West, these years were marked by vertiginous social and economic development. It was a period characterized by continual growth sustained by unlimited access to fossil fuel resources like petroleum and uninterrupted technological and scientific progress. US president Harry Truman summarized this euphoric confidence in the future and progress in his inaugural address of January 20, 1949, when he said, "Greater production is the key to prosperity and peace. And the key to greater production is a wider and more vigorous application of modern scientific and technological knowledge."[3]

Faith in scientific and technical progress, atomic energy research and space exploration – with their military implications – paradoxically overlapped a new ecological awareness. July 16, 1945, the date the first atomic bomb was exploded in the New Mexico desert, also heralded the dawn of a new era of concern for the environment. And yet it was the satellites, Russian and American missions (particularly the Apollo missions), and space stations orbiting the earth that made it possible to perceive our planet as a coherent whole and comprehend it in relationship to the other celestial bodies. In the missions leading up to Apollo 11, public interest was already divided between discovery of the moon and rediscovery of the earth. In the case of the Apollo 8 mission in December 1968, NASA's press review for Christmas Eve included thirteen front pages of American newspapers with photos of the earth seen from space, five with photos of the moon and five with both.[4]

The images of the earth captured from space fostered a popular shared notion of a "Spaceship Earth," for which Buckminster Fuller wrote an *operating manual*.[5] Only through these views of our planet was it possible to associate an image with the idea of the earth as an ecosystem, a unique space, limited and precariously balanced. NASA itself sought to redeem or minimize its involvement in the military field by conducting research on the earth's atmosphere, the condition of the ozone layer and the process of desertification.[6] The exploration of space in the 1960s and 1970s therefore did not embody the idea of conquest represented, for example, in the words of John F Kennedy ("It will not be one man going to the moon… it will be an entire nation"[7]), nor did it promise a future without limits or barriers in which to project the idea of progress and continual growth predicted by Truman. Paradoxically, it contributed to the rediscovery of the earth and its limitations.

American astronaut Edwin "Buzz" Aldrin and Tuscan peasant Zeno Fiaschi were main figures in *Architettura interplanetaria* (Interplanetary architecture) and *Cultura materiale extraurbana* (Extra-urban material culture), both projects by Superstudio and Alessandro Poli. Of all their projects, these are the most extreme. They explore, on one hand, a new absolute landscape free of earth's political, social and cultural constraints but at the same time rife with new limitations and constraints, and on the other hand, a rural culture that is becoming extinct. The astronaut Aldrin and the peasant Zeno lived in these landscapes, moving along the highways of space in one case and within fragile huts in the Tuscan countryside in the other. Both relied on a panoply of objects, or rather tools, for their survival: for Aldrin, primarily the space suit; for Zeno, tools reinvented and adapted to the most diverse agricultural and everyday uses. Both sets of tools were the result of a sophisticated way of thinking that requires optimal use of the few available resources.

Edwin Aldrin returned from the Apollo mission in 1969. Zeno Fiaschi died in 1982. In the late 1990s, in one of his conversations with David Cayley, Ivan Illich addressed the topic of technology and tools. In accordance with the intuitions of philosopher Carl Mitcham, Illich observed that the idea of the tool had not always existed but originated between the thirteenth and fifteenth centuries. He suggested that "sometime during the 1980s the technological society which began in the fourteenth century came to an end… It appears to me that the age of tools has now given way to the age of systems, exemplified in the conception of the earth as an ecosystem, and the human being as an immune system."[8]

Other Space Odysseys has nothing to do with Space Architecture or architecture in outer space. It is not a heroic celebration of tools, high-tech architecture and imagery, extreme physical (or mental) conditions, or advanced technology meant to sustain human beings in such an environment. Our renewed interest in space is not prompted by the banalization of space brought about by new efforts to exploit it commercially (communications, weather forecasting and even energy production) or to develop space tourism. Space tourism may in fact be considered the extension into space of the tourism that has assailed the entire planet and contributed, along with urbanization, to the spread of communications networks, the mediatization of events and the creation of an urban world that has subsumed various cultures, as Superstudio observed in the introduction to its study of material culture (and of Zeno). In this new world, we are witnessing the triumph of global consumption, of which tourism is the most representative expression, and migration the "hidden" facet. Late twentieth-century architecture has represented these two aspects of this contemporary reality, on one hand through the celebration of the technocratic features of high-tech architecture – increasingly cloaked in ecological justifications – and on the other through the creation of new monuments of the global economy and the media society, "planetary curiosities with the power to attract an influx of tourists from all over the world." [9]

Using the moon landing as his starting point, Alessandro Poli introduced us to a new landscape in a letter to Adolfo Natalini (which he never sent). [10] However, the exploration of space was above all a television event, which produced a true, new, "radical environment": the global media landscape. Today, space can again offer us the possibility of thinking about a new landscape, just as it did for Superstudio in the 1970s: a theoretical landscape in which to approach the question of constructing a totally artificial world free from gravity, as with Greg Lynn; or to rethink the techno-bureaucratic imagination that pervades contemporary society through an architecture conceived as a space of the mind, as with Michael Maltzan.

The exploration of space not only produced the apotheosis of the concept of the tool, and its spinoff on earth in the hands of the peasant Zeno. Space can show us a new world where not only would objects be forgotten but the tools of Aldrin and Zeno would also be abandoned, carrying us into the realm of systems foretold by Ivan Illich. Once again, as Superstudio wished, it is a question of letting go of architecture understood as the production of material goods in favour of architecture as the production of ideas.

1   Gabriel Tortella, *The Origins of the Twenty-First Century: An Essay on Contemporary Social and Economic History*, trans. M Carmen Fayos de Riddel, ed. Michele Schiavone (New York: Routledge, 2010), 144. Originally published as *Los orígenes del siglo XXI. Un ensayo de historia social y económica* (Madrid: Gadir, 2005).

2   This expression was introduced by Jean Fourastié in *Les Trente Glorieuses ou La révolution invisible de 1946 à 1975* (Paris: Le livre de poche, 1989).

3   Harry S Truman, inaugural address, January 20, 1949, Washington DC, in *Fellow Citizens: The Penguin Book of U.S. Presidential Addresses* (New York: Penguin Books, 2008), 365.

4   Robert Poole, *Earthrise: How Man First Saw the Earth* (London and New Haven: Yale Univ. Press, 2008), 31.

5   R Buckminster Fuller, *Operating Manual for Spaceship Earth* (Carbondale: Southern Illinois Univ. Press, 1969). New edition published in 2008 by Lars Müller, Baden.

6   Joachim Radkau, *Nature and Power: A Global History of the Environment*, trans. Thomas Dunlap (Cambridge: Cambridge Univ. Press; Washington, DC: German Historical Institute, 2008), 291. Originally published as *Natur und Macht. Eine Weltgeschichte der Umwelt* (Munich: C H Beck, 2000).

7   Robert Friedel, *A Culture of Improvement: Technology and the Western Millennium* (Cambridge, MA: MIT Press), 528.

8   Ivan Illich, *The Rivers North of the Future: The Testament of Ivan Illich*, as told to David Cayley (Toronto: Anansi, 2005), 77. This was a new concept for Illich. As he noted in *Rivers North*, in previous works such as *Tools for Conviviality* (New York: Harper & Row, 1973), he considered tools as a permanent concept.

9   The original quote is "… des curiosités planétaires susceptibles d'attirer les flux touristiques mondiaux." Marc Augé, *Où est passé l'avenir?* (Paris: Panama, 2008), 66.

10   Please see pp. 84–85 in the present volume.

Fig. I.

Fig. II.

Fig. III.

Fig. I	A. Mare degli Umori	5 . Gassendi
PIANETA MARTE	B. „ ' delle Nubi	6 . Schickard
	C. Oceano delle Procelle	7 . Arpalo
Fig. II	D. Mare di Nettare	8 . Eraclide
GIOVE	E. „ della Tranquillità	9 . Landsberg
	F. „ della Serenità	10. Reinhold
Fig. III	G. „ della Fecondità	11. Copernico
SATURNO	H. „ delle Crisi	12. Elicona
	1 . Grimaldi	13. Capuano
	2 . Galileo	14. Bulialdo
	3 . Aristarco	15. Eratostene
	4 . Kepplero	16. Timocari

Telescopic representations of some
of the most important objects of the sky
Aquatint etching, 19th century

PORTANTI OGGETTI DEL CIELO

29. Plinio
30. Teofilo *(Caterina Cirillo)*
31. Fracastoro
32. Censorino
33. Messala
34. Fromont. e Palude del Sonno
35. Proclo
36. Cleomede
37. Snellio e Funerio
38. Petavio
39. Langreno
40. Terunzio

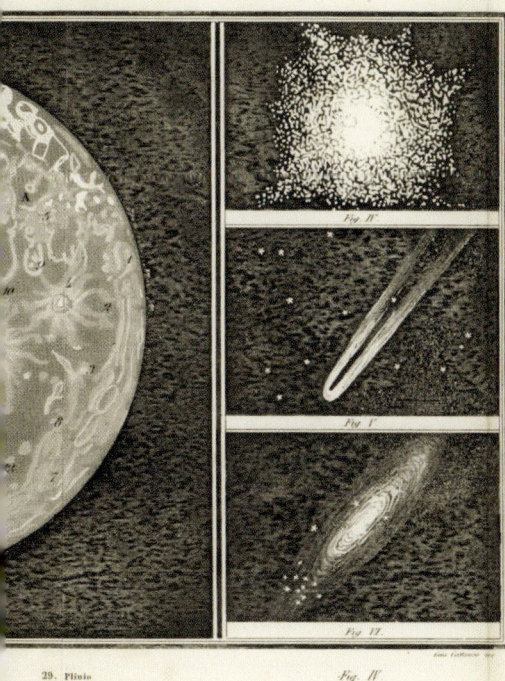

Fig. IV.

COACERVAZIONE STELLARE

Fig. V

COMETA

Fig. VI

NEBULOSA

Acknowledgements

This book and the exhibition it accompanies are part of the Canadian Centre for Architecture's ongoing investigation of architecture's role in light of current economic, social, technological and environmental issues.

Other Space Odysseys could not have been produced without the collaboration of a great number of individuals who contributed their skills, advice, and support throughout the project.

First and foremost, we would like to acknowledge the role of CCA Founding Director Phyllis Lambert, who supported the effort from the outset and recognized the exhibition's potential for advancing the debate on architecture as a way of producing new ideas.

We would like to thank Greg Lynn, Michael Maltzan and Alessandro Poli for their engagement in this project: they contributed with new concepts and advanced research in their works, and they welcomed us into their offices, archives and homes, allowing us to realize this project as an intriguing dialogue between them and the CCA on the specific topic we have chosen. Special thanks go to all the members of their offices for their invaluable contributions in the development of the book and the exhibition: in particular we would like to thank Greg Lynn FORM, Jackilin Bloom, Micael Duran, Eric Leishman, Lisa Sommerhuber and from Michael Maltzan Architecture, Mirella Abounayan, Wil Carson, Stacie Escario, Christopher Norman and Tim Williams.

We are grateful to Lars Müller for his work on this volume, for his brilliant interpretation and understanding of its content and its scope. We would like thank Tessy Ruppert for her careful graphic design work.

Special thanks go to Alex DeArmond for bringing his sophisticated insights to the graphic design of the exhibition.

We would also like to express our greatest appreciation to Nargisse Rafik and Katya Epstein for their invaluable work in the editorial shaping of the texts compiled here, and to Lev Bratishenko and Peter Sealy for their skilled research and ingenuity, which helped us to clarify some of the contributions in this volume. Special thanks also to the translators, Albert Beaudry, Donald Pistolesi, and Dominic Brierre, for interpreting the contents of this book with care and fidelity, and to Isabelle Canarelli, Katie Moore and Trish O'Reilly for their thorough work in copyediting and proofreading it.

Many thanks go to the entire CCA team for its dedication to the realization of the book and the exhibition in Montréal. In particular, we would like to thank Daria Der Kaloustian for her invaluable contribution in directing the project; Meredith Carruthers whose thorough editorial assistance, research and coordination skills made possible a project with a dispersed team of international collaborators and contributors working from Los Angeles, Minneapolis, Montréal, Milan, Florence and Baden; Laura Killam and Sébastien Larivière for their excellent work on the realization of the exhibition design proposed by Lynn, Maltzan and Poli; and Elspeth Cowell for confidently handling many of the administrative aspects of the project. Our gratitude also goes to the CCA Library team for their invaluable assistance during the research phase of this project and to the CCA conservators for their expertise in handling a display that proposed an unusual way of presenting original materials in a museum environment.

Finally, we would like to express our appreciation to Sina Najafi, with whom we had many extensive conversations in New York and Montréal.

Mirko Zardini, Director and Chief Curator
Giovanna Borasi, Curator of Contemporary Architecture
Canadian Centre for Architecture

Copyrights and Photo Credits

Projects

N.O.A.H. (New Outer Atmospheric Habitat) Sets for the Film *Divide*, 2004
Jörg Tittel and Ethan Ryker (film authors)
Greg Lynn FORM, Los Angeles, CA (N.O.A.H. design)

New City (Commissioned for *Design and the Elastic Mind* exhibition at the Museum of Modern Art), 2008–
Greg Lynn, Peter Frankfurt (with Imaginary Forces) and Alex McDowell

Space Studio, University of California, Los Angeles, 2007–2008 and University of Applied Arts Vienna, 2006–2007

Jet Propulsion Lab, a new building proposal, 2010–
Michael Maltzan (Design Principal), Peter Erni (Project Director), Kristina Loock (Project Architect), Wil Carson (Project Designer), Sahaja Aram (Job Captain), Sevak Karabachian, Michael Leaveck, Christian Nakarado, Yan Wang (Project Team)

Architettura interplanetaria, 1970–71
Superstudio (Adolfo Natalini, Cristiano Toraldo di Francia, Roberto Magris, Gian Piero Frassinelli, Alessandro Magris, Alessandro Poli [1970–72])

Piper, 1966
R Gherardi, S Pacini, A Poli, R Russo, F Spinelli

Zeno: Research on a Self-Sufficient Culture, 1979–80
Alessandro Poli

This volume is published by the Canadian Centre for Architecture and Lars Müller Publishers in conjunction with the exhibition *Other Space Odysseys: Greg Lynn, Michael Maltzan, Alessandro Poli*, organized by the Canadian Centre for Architecture, Montréal, and presented at the CCA from 8 April to 6 September 2010. Issued also in French under the title *Autres odyssées de l'espace : Greg Lynn, Michael Maltzan, Alessandro Poli*.

Printed in Germany
First edition
Printed on acid-free paper

978-0-920785-88-1
Canadian Centre for Architecture
1920 rue Baile
Montréal, Québec
Canada H3H 2S6
www.cca.qc.ca

978-3-03778-193-7
Lars Müller Publishers
5400 Baden, Switzerland
www.lars-mueller-publishers.com

Bibliothèque et Archives nationales du Québec and Library and Archives Canada cataloguing in publication
Other space odysseys : Greg Lynn, Michael Maltzan, Alessandro Poli
(Manifesto)
Catalogue of an exhibition held at the Canadian Centre for Architecture, Montréal, Québec, Apr. 8-Sept. 6, 2010.
Issued also in French under title: *Autres odyssées de l'espace*.
Co-published by Lars Müller.
ISBN 978-0-920785-88-1
1. Visionary architecture – Exhibitions. 2. Lynn, Greg, 1964– – Exhibitions. 3. Maltzan, Michael – Exhibitions. 4. Poli, Alessandro, 1941– – Exhibitions. I. Borasi, Giovanna, 1971– . II. Zardini, Mirko. III. Lynn, Greg, 1964– . IV. Maltzan, Michael. V. Poli, Alessandro, 1941– . VI. Canadian Centre for Architecture.
NA209.5.O84 2010   720.1   C2010-940652-4

Publication
Editors: Giovanna Borasi and Mirko Zardini
Editorial assistant and project coordination: Meredith Carruthers
Rights & Reproductions: Elspeth Cowell
Editing: Katya Epstein
Translation: Donald Pistolesi
Design: Integral Lars Müller, Lars Müller and Tessy Ruppert

Exhibition
Curators: Giovanna Borasi and Mirko Zardini
Concept: Greg Lynn, Michael Maltzan, and Alessandro Poli
Research and project coordination: Meredith Carruthers
Design Development: Laura Killam and Sébastien Larivière
Graphic Design: Alex DeArmond

The CCA is an international research centre and museum founded on the conviction that architecture is a public concern. Based on its extensive collection, exhibitions, programs and research opportunities the CCA is a leading voice in advancing knowledge, promoting public understanding, and widening thought and debate on the art of architecture, its history, theory, practice, and role in society today.

The exhibition catalogue is prepared in part, thanks to the financial support of Hydro-Québec. The CCA would also like to thank the Ministère de la Culture, des Communications et de la Condition féminine, the Canada Council for the Arts, and the Conseil des arts de Montréal for their continuous support.